WHERE THE

WILD THINGS

WERE

WITHDRAWN

WHERE THE WILD THINGS WERE

Life, Death, and Ecological Wreckage

in a Land of Vanishing Predators

William Stolzenburg

BLOOMSBURY

NEW YORK · LONDON · NEW DELHI · SYDNEY

Copyright © 2008 by William Stolzenburg

All rights reserved. No part of this book may be used or reproduced in any manner whatsoever without written permission from the publisher except in the case of brief quotations embodied in critical articles or reviews. For information address Bloomsbury USA, 175 Fifth Avenue, New York, NY 10010.

Two lines from "The Bloody Sire" from *The Selected Poetry of Robinson Jeffers*, copyright by the Jeffers Literary Properties. All rights reserved. Used with the permission of Stanford University Press, www.sup.org.

Published by Bloomsbury USA, New York

All papers used by Bloomsbury USA are natural, recyclable products made from wood grown in well-managed forests. The manufacturing processes conform to the environmental regulations of the country of origin.

LIBRARY OF CONGRESS CATALOGING-IN-PUBLICATION DATA

Stolzenberg, William.
    Where the wild things were : life, death, and ecological wreckage in a land of vanishing predators / William Stolzenberg.—1st U.S. ed.
        p.   cm.
    Includes bibliographical references.
    ISBN-13: 978-1-59691-299-1 (hardcover)
    ISBN-10: 1-59691-299-5 (hardcover)
        1. Predation (Biology) 2. Predatory animals. 3. Conservation biology.
    4. Endangered ecosystems. I. Title.

    QL758.S746 2008
    577'.16—dc22
    2008002392

First published by Bloomsbury USA in 2008
This paperback edition published in 2009

Paperback ISBN-10: 1-59691-624-9
ISBN-13: 978-1-59691-624-1

    3 5 7 9 10 8 6 4

Typeset by Westchester Book Group
Printed in the United States of America by Quebecor World Fairfield

ACC LIBRARY SERVICES AUSTIN, TX

To Mom and Dad

To Mom and Dad

CONTENTS

PROLOGUE: The Grizzly in the Room

ANYONE WHO WRITES a book of science about great, flesh-eating beasts should be required up front to disclose their bias. Here is mine.

The second week of June 2000, on the campus of the University of Montana in Missoula, nearly a thousand professional biologists and advanced students had gathered for the fourteenth annual meeting of the Society for Conservation Biology. As a science writer covering that conference, I was to navigate the chaos of some four hundred presentations—going off eight at a time, thirty-two to the hour, three days straight, in various locations about campus. My reporting strategy, honed to questionable success with years of practice, was to scramble hither and yon in manic pursuit of the most captivating titles, the next great thing in conservation, as one might try fielding an exploding batch of popcorn.

Within fifteen minutes of the first day's opening sessions, that strategy had been scrapped for an infinitely more alluring one. I had found a seat in a symposium called "The Role of Top Predators in Ecological Communities and Biological Conservation," and for the next three hours I made no pretenses of needing to be anywhere else.

Because here were stories of lions, hyenas, and leopards, battling cheetahs and wild dogs over bloody carcasses on the African plains. Here were tales of wolves raising the neck hairs on moose in Alaska, rogue killer whales gobbling sea otters in the Aleutians, even coyotes chasing house cats in California suburbs. Here, one after the next, were legitimately visceral alternatives to filing yet another pale report on habitat fragmentation, population viability analyses, or microsatellite loci.

The faces materializing at the podium, the names appearing on the papers, included some of the icons and iconoclasts of conservation biology. There was James A. Estes, pioneering marine biologist whose observations in the Aleutians thirty years earlier had revealed the sea otter as resurrected guardian angel of the vibrant Pacific kelp forest. Estes was back from yet another season in the cold northern waters, with a bizarre new twist in the otter's comeback story—a twist involving otter-eating killer whales, whose punch line still has the marine science community feuding.

Estes also brought a message from his coauthor John Terborgh, a legendary tropical ecologist with nightmarish news from a surreal, predator-free archipelago in Venezuela, whose forests in the absence of harpy eagles and jaguars were being eaten to the ground. There was Joel Berger, noted authority on large hoofed animals, diagnosing a strange case of amnesia in the Grand Teton range of Wyoming, where the moose had forgotten certain essential fears—an unfortunately lethal lapse now that wolves had recently reclaimed lost ground in the Tetons.

Later that afternoon, Berger's colleague Peter Stacey was revealing more wounds of missing predators. Streamside birds of the Grand Teton had disappeared, in a chain reaction eventually traced back to the mountains' missing wolves and grizzlies.

The next day brought more on the science of predators and predation as increasingly vital matters in conserving life's diversity. A progress report from Yellowstone National Park, then five years into a bold experiment of turning gray wolves loose after a seventy-year hiatus, suggested the sanctuary had been decidedly shortchanged in the wolves' absence. The reinstated top predator, reported lead researcher Douglas Smith, was turning the park into a banquet of elk carrion, with a slew of scavenging species reaping the leftovers. It would turn out these were the rumblings of bigger tremors to come; Smith and his colleagues were sitting cautiously on preliminary findings of a wholesale revival of Yellowstone's compromised ecosystem, courtesy of the wolf.

Talk after talk, northern seas to tropical jungles, the conclusions rang in accord, as with a gavel: Big predators were not just missing; they were sorely missed. It brought to mind a medical phenomenon haunting many amputees; the phantom pains of a missing limb. These top predators—these missing limbs—were still deeply felt.

Here, in a country whose society had blown away all but a token remnant of its topmost competitors, was a force of top-flight ecologists exposing the campaign as a colossal case of shooting one's own foot. Here was evidence that the biggest and scariest of carnivores were more dangerous by their absence. It was time, as Jim Estes addressed his audience, to rethink the way we look at the world, to consider the view from the top down—from the predators' perspective.

From that day I began tracking this insurgent cadre of concerned scientists taking stock of Earth's increasingly fangless kingdom. Theirs is the story of this book. In field sites spanning the biosphere, these ecologists are questioning the soundness of ecosystems recently devoid of their topmost predators, and discovering suspicious cracks in the foundation. They are flagging, in a sense, what the bard of ecology Aldo Leopold once described as "the marks of death in a community that believes itself well." And I hope, if nothing else, that through the following chronicle of their discoveries, these unseen wounds and phantom pains, whose source scientists are now bringing to light, may at least be made visible for all of us to deal with as we choose.

But again to that sticky business of bias. There is a reason these discoveries have been so late in coming and—as we'll see—so warily received. The ecology of big predators remains the most intractable discipline in the most complex of all sciences. Its subjects are hard to find, and harder yet to hold still for study. The big predators are not only inherently rare—as ordained by their tiny perch atop the food pyramid—but fashionably rare, at the hands of a modern human society that slaughters them blatantly out of contempt and obliquely through wholesale destruction of their homes and livelihoods. These are animals that tend to roam too far for conventional observation, considering that a week's jaunt by a lovelorn wolf may span entire western states. The great carnivores, like lions melting into the tall grass, are also by nature enigmatic and stealthy, and dangerous when cornered.

Their intimate study poses logistical and psychological issues unknown to the biologist studying deer or beetles.

And therein lies one confounding variable that inescapably pervades this supposed book of fact and systematic inquiry. Over the thousands of millennia that our own lineage has spent in the company of killing beasts—competing with them for food and running from them as food—the great meat-eaters have quite naturally etched themselves into the human persona. Long before people had perfected the art of exterminating their fellow predators, they were worshipping them. Thirty thousand years ago, Paleolithic artists were decorating cave walls with reverently painted murals of lions. To this day no human, scientist or otherwise, impassively witnesses the disembowelment of a living creature by the tearing jaws and claws of wild carnivores. No one impartially records the soul-jarring charge of a grizzly bear or the mountain-hushing howl of the wolf.

The question hanging over that Missoula gathering was palpable, and it has stalked this story to the end. All this talk of killing and fleeing and ecological chain reactions made for stirring copy, but was it legitimate? Were these the reports of sound science or the veiled advocacy of a few who had fallen prey to the predators' mystique? More to the heart, what would it mean to the human animal to one day wake up and find itself in a world where the biggest, most threatening predator in the whole blessed menagerie was a coyote the size of a border collie? These concerns were pertinent not only to the predator ecologist advocating conservation but also to the journalist who would question their conclusions. Something more than science pervaded the discussions—something akin to the five-hundred-pound grizzly in the room.

Nine days after my awakening in Missoula, I was standing alone at dusk just outside the northwest corner of Yellowstone National Park, atop an open knoll ringed on all horizons by ranges of the northern Rockies. Earlier in the day and far behind I had passed a sign at the trailhead reading WARNING: GRIZZLY BEARS ARE ACTIVE IN THIS AREA. And after reconnoitering the countryside, I had determined that this natural amphitheater would indeed be a good place to seek an answer or two about the objective nature of carnivore journalism.

As the skylight faded I stood on the lonely knoll, slowly turning in circles. There was less faith than duty in my exercise, scanning the surrounding hillsides for bears I held no serious hopes of conjuring. Around I turned, drifting between distant mountain peaks and the purpling skies in the purest of silence. Duskdreaming. The trail heading back would soon be too dark to follow. Time to go. I turned one last quarter to the north, and there stood grizzlies.

A sow and two cubs had magically materialized on the hillside, two roundish nubs trailing behind a dark boulder of fur, placidly pawing through a seep on the edge of an aspen grove. I slowly raised my binoculars. I lowered them and looked to the lone pine standing about a seven-second sprint to my left, estimating its lowest branch at six feet high. I pulled out my field journal and scribbled some notes, ostensibly recording some key facet of natural history I pretended to be observing.

When I look at those notes today I see the jerky scratchings of an overexcited child. When I remember the heart-pounding presence those bears had imparted over the distance—a distance that could just as likely have measured a hundred paces as a half mile—I remember why this book, for all its inherent hazards, needed writing.

ONE Arms of the Starfish

ON THE NORTHERNMOST TIP of Washington's Olympic Peninsula, in a wild and lonely little crescent of shore called Mukkaw Bay, ocean meets land in a crash of wind and wave against craggy rock, geysers of salt spray erupting into brooding skies. Mukkaw Bay is the sort of place to imagine the Romantic philosopher, hair swept crazy by the gales, pondering the epic clash of titanic forces. And so in a sense it came to be, when in the summer of 1963 a young farsighted ecologist happened upon the place with a few pressing curiosities regarding the vital struggle between predator and prey.

It was thereafter that the austere seascape of Mukkaw Bay came to be periodically enlivened by the spectacle of a tall figure scrambling about on the fleetingly exposed terrain of low tide, bounding on long legs across the tidal pools and crevices and surge channels, stepping nimbly over slippery strands of glistening kelp. Each time, arriving at the same particular ledge, the young man would stoop to pry something from the rocks and, one by one, begin flinging starfish into the sea.

Robert T. Paine was a freshly minted ecology professor from the University of Washington in Seattle, with questions about things as fundamental as a predator's impact on its prey, as grand as the principles underlying the diversity of life. On the rocky shores of Mukkaw Bay, he had come looking for answers in a colorful little community of marine invertebrates—barnacles, limpets, snails, mussels, chitons, and starfish—conveniently gathered, readily manipulated, and ripe for experimentation. Paine set about sampling two adjacent intertidal stretches of rock on the wave-beaten edge of the sea.

Each month he would make his way out to the turbulent ledge, and from one of his plots he would meticulously chuck back into the ocean every last individual of the ecosystem's reigning predator, a husky orange starfish formally known as Pisaster ochraceous. The other rock he left untouched.

Paine did not need to stare long to decipher the outcome. While the untampered plot had continued merrily along—with its cast of characters fully intact—its predator-free counterpart next door had fallen under siege. Where the predator Pisaster went missing, its main prey, a big dark mussel named Mytilus californianus, flourished spectacularly. Within a year, Mytilus had crowded half the other species off the rock, with the survivors hanging on by their figurative fingertips. In time, only a stark monoculture of mussels would remain. As orchestrator of Pisaster's local extinction, Paine had triggered the collapse of his miniature ecosystem.

Robert T. Paine's muscular little experiment, published in a 1966 issue of the scientific journal the American Naturalist, was to become a classic paper in the nascent field of community ecology. The most elementary interpretation was deceptively powerful: On this little stretch of rocky shore, one particular kind of animal wielded a disproportionately huge hand in determining how many species shared the rock. Reading more deeply, the implications of Paine's predator-free play world grew quietly dark and monstrous.

Spitsbergen
Paine's inspired path to Mukkaw Bay had begun with another young man's odyssey nearly half a century before and a continent away, but in the vaguely familiar setting of a wild and weather-beaten land beside the sea. The man was Charles Sutherland Elton, a naturalist prodigy who in 1921, as a twenty-one-year-old zoology student at Oxford, was offered a slot on the university's first Arctic expedition. The land was Spitsbergen, an ice-encrusted, polar-bear-haunted archipelago the size of Ireland, anchored in the Barents Sea roughly midway between Norway and the North Pole. Spitsbergen—German for "spiked mountains"—was a beautifully bleak and treeless tundra, more than half of it covered year-round in ice. Isolated from the mainland by hundreds of miles of Arctic waters, previously scoured by the grindings of Pleistocene

glaciers, Spitsbergen was home to a paltry but importantly conspicuous little collection of wildlife castaways. Not including the polar bears and walruses and seals that made their living upon the sea, the mammalian fauna of terrestrial Spitsbergen numbered all of two species, the reindeer and a fox. The bird list was likewise threadbare: a handful of part-time resident seabird species—albeit in teeming, cacophonous colonies—and on land a snow bunting, a lone species of sandpiper, and a snow-white tundra grouse named the Svalbard ptarmigan. Aside from some insects and spiders to round out the collection, that was it. No lemmings or Arctic hares, not much of anything but the barest-boned collection of the north's hardiest hangers-on—which was, after all, the main allure of Spitsbergen.

In those adolescent days of ecology, biological surveys were in vogue. The ecologists of the time had progressed from naming and cataloging individual species of plants and animals to attempting the same treatment for the communities in which the species lived. In these early stumbling steps, the simpler the better. "This poverty in species," Elton later wrote of Spitsbergen, "made it possible to carry out a fairly good primary ecological survey in spite of the inaccessible nature of much of the country."

Young Elton came ready. Since the age of seventeen, Elton had been staring out at the passing countryside from the windows of railway trains, walking spellbound through the English woods, quietly chasing "the dream of really knowing someday what animal populations are doing behind the curtain of cover." He was a field naturalist of the old school, fluent in the names and habits of birds, mammals, insects, and rocks alike, who could be found lying on his belly deciphering the complexity of a pond, or hunched over an anthill in the English countryside, mesmerized by the commerce of its creatures. His days were spent noting robber ants plundering, bumblebees fertilizing thyme flowers, rabbits depositing dung, and green woodpeckers digging. Elton early on had assumed the task of looking into the clandestine society of creatures whose very nature it was to hide. At least on Spitsbergen, the hiding places were few.

At the base of the great sea cliffs of Spitsbergen, the islands' dull brown tundra suddenly took on the emerald hues reminiscent of the poster hills of Norway. (Elton didn't offer this comparison blithely; he tested it with a color chart.) Beneath the cliffs grew luxuriant swards of grass and flower gardens of saxifrage. This apparent oasis sprouting from the tundra bleakness was no microclimatic fluke of temperature or rainfall. The source of this greenery, Elton plainly realized by looking up, sprang from guano.

The cliff faces of Spitsbergen were plastered with raucous seabird colonies of guillemots and puffins, a controlled riot of uncountable thousands. Beneath these bustling hubs of birdlife, the guano rained, and the coastal gardens grew lush. In the everlasting daylight of the Arctic summer, the seabirds were forever commuting in a continuous mad traffic, from sea cliff to sea and back again. Elton's mind followed them out of sight into the cold gray waters. Far out to sea on egg-beater wings they would fly to their feeding grounds. Then into the water they would dive, becoming nimble submarine predators—web-footed torpedoes chasing down scores of little fish and shrimplike krill. Beaks and stomachs full, they would fly back to their cliffside nests bringing food, in one state of digestion or another, to hungry seabird chicks and tundra plants alike.

Watching this endless influx of wildflower fertilizer flown in from the sea, Elton saw a chain of life far transcending all preconceived bounds of the animal community. He saw it beginning with the base of the oceanic food chain—with the infinite masses of photosynthetic plankton called diatoms—feeding great planktonic herds of diatom grazers, they in turn feeding the little fish and krill destined for the beaks of landward-bound seabirds. Inland of the tundra wildflowers, he saw the chain stretching even further, to insects and spiders, to the beaks of buntings and the jaws of foxes. It was more than a chain of food; it was of web of interactions, ultimately transforming the face of the land. The pastures of the sea were fertilizing terrestrial gardens. And animals were doing much of the heavy lifting.

Everywhere he turned on Spitsbergen, Elton saw the horizons shifting. Which was also to say, everywhere he turned, it seemed, he saw the Arctic fox. Chief rascal of the High Arctic, *Alopex lagopus* was forever crossing Elton's path in its peripatetic search for food. The fox could be found anywhere, chasing ptarmigan and snow buntings in the mountainside tundra, nabbing fallen seabirds beneath the coastal cliffs. Once, while the expedition was off exploring the Nordenskiöld glacier, hungry foxes sniffed out the researchers' cache of rock and fossil specimens—all dutifully labeled—and made food of the labels.

As a hunter, the Arctic fox followed its nose to the ends of Spitsbergen, and then much farther. With descent into the eternal twilight and deep freeze of the Arctic winter, with the seabirds long departed and the ptarmigan tunneled beneath the snow, the fox was known to follow its hunger out onto the pack ice, chasing polar bears. There it lived for months on the bear's leavings, gnawing scraps of seal carcass, noshing piles of bear dung. Then back it returned to Spitsbergen with the vernal sun, changing its uniform from winter white to summer brown, and its profession from sea-ice scavenger to the tundra's top-rung predator. The Arctic fox had slipped through yet another hole in the tidy conceptual fence surrounding the animal community.

In Spitsbergen, Elton could see forever. In this supposedly simple collection of Arctic castaways, there was a complex and far-flung commerce among plant and animal, of species interacting across the distances, tipping balances, triggering chain reactions. Here, in Elton's words, was "a very untidy, dynamic, mobile, changing picture of nature rather than a neat physiological arrangement; a world of rather unstable populations in an unstable environment, not a static arrangement of animals limited to habitats created chiefly by plants and vegetation with its special microclimates."

This was a land where animals were no longer pawns, but players. To define the ecological community by simply listing the most conspicuous or characteristic inhabitants, was, to Elton, "to put nature into a physiological strait-jacket." It was to overlook the work being done, the muscle involved, "the main cogs in the *heavy machinery* of the place."

It has been suggested that Spitsbergen had served Elton as the Galápagos had served Darwin. In their surface simplicity, both lands had parted the clouds for two remarkably receptive and penetrating minds. Through the naked rubble of the Galápagos Archipelago, a young Charles Darwin had seen the struggles for survival, and in them the seeds of his earth-shaking explanation for the divergence and origin of species; in the ice-bound expanses of Spitsbergen, Charles Elton had caught a glimpse of the hidden glue that held them all together.

In 1927, his mind straining at the seams from three summers at Spitsbergen, Elton sat down at his desk and released the floodgates. In a ninety-day burst of creative fervor he wrote a two-hundred-page book, published simply as *Animal Ecology*. Eighty years later, the elegant little volume remains a standard on reading lists and bookshelves of students and professors of ecology.

In clear bold tones and basic English, *Animal Ecology* frames the questions, and a good many of the answers, that still occupy the major discussions of modern science. Gems of Elton's prescience can be found on nearly every page of *Animal Ecology*. But one could start by flipping to chapter 5, "The Animal Community," and immediately find the essence of Elton's ecological perspective. There at the head of the page it is laid out in three Chinese proverbs:

"The large fish eat the small fish; the small fish eat the water insects; the water insects eat plants and mud."

"Large fowl cannot eat small grain."

"One hill cannot shelter two tigers."

This was Elton's way of introducing several of the most important concepts in the field of community ecology. It is a chapter that begins with Elton studiously watching an anthill and finishes with a lion killing zebras. And in between those poles, he boils down the whole of animal society to a word, food:

Animals are not always struggling for existence, but when they do begin, they spend the greater part of their lives eating . . . Food is the burning question in animal society, and the whole structure and activities of the community are dependent upon questions of food-supply.

Elton's *Animal Ecology* draws heavily on his apprenticeship in Spitsbergen. He had ultimately discovered order in the freewheeling Arctic assemblage, and it had come to his mind in familiar shapes and constructions. He saw the sun's energy linked to the greenery of tundra plants to the feathers of ptarmigan to the muscle of Arctic fox. It was simple enough: Plants eat sunlight, herbivores eat plants, carnivores eat herbivores, "and so on," he wrote, "until we reach an animal which has no enemies." Life was linked in chains of food.

One could find a welter of such chains, even in the skeleton crew of species that epitomized Spitsbergen. Here Elton included a hand-drawn sketch. In what otherwise resembles the electrical diagram of a circuit board, Elton's lines and arrows run this way and that, connecting boxes labeled guillemots and protozoa, dung, spider, plants, worms, geese, purple sandpiper and ptarmigan, mites, moss, seals, polar bear, more dung, more arrows, all arrows eventually leading to the Arctic fox.

The chains are intertwined, crisscrossing and connecting, forming what Elton had come to think of as a "food-cycle," what his descendants today call a food web. Elton, in his fascination with numbers, started adding them to that web: Roughly how many plants and plankton at this end of the chain, how many foxes and bears at that end? With the numbers, his web had gained a third dimension, and its shape now took the form of a pyramid.

Elton's pyramid is a narrowing progression in this community of life, founded on a broad, numerous base of plants and photosynthetic plankton—harvesters of the sun's energy, primary producers of food. From there it steps up to a substantially more narrow layer of herbivorous animals cropping their share from below, and so on up to yet a smaller tier of carnivores feeding on the plant-eaters. Perched loftily at the apex are the biggest, rarest, topmost predators, those capable of eating all, and typically eaten by none. In the prolific plains of the Serengeti, that predator would be the African lion; in the stingy tundra of Spitsbergen, that predator happens to be a scrappy little fox, "the apex of the whole terrestrial ecological pyramid in the Arctic." Elton's geometric perspective on life would soon become one of the tenets of ecology, and to this day is known as the Eltonian pyramid.

Animal Ecology is where the food chains that Elton realized in the guano gardens of Spitsbergen, where the pyramid of numbers he saw spreading beneath the Arctic fox, are set down as principles for all life on Earth.

Along the way, Elton also discussed the importance of body size with re-

gard to the act of eating and being eaten—stating again the obvious observation whose importance had somehow escaped so many before him.
"There are very definite limits, both upper and lower, to the size of food
which a carnivorous animal can eat . . . Spiders do not catch elephants in
their webs, nor do water scorpions prey on geese." He also gave new life
and meaning to the word "niche," through his own definition: An animal's
"place in the biotic environment, *its relations to food and enemies*."

The Chinese may have offered the original inklings on animal ecology,
but it was Elton who built a pyramid out of them. With his niches, his food
chains, his pyramids, Elton gave his fellow ecologists homework assign-
ments for the coming century.

The Struggle for Existence

It was easy enough to see, with the help of Elton's timely reminder, that life
was stacked in a pyramid of numbers. But what controlled those numbers?
"What prevents the animals from completely destroying the vegetable and
possibly other parts of the landscape," asked Elton. "That is, what preserves
the balance of numbers among them (uneasy balance though it may be)?"

In the 1930s, a Russian microbiologist named Georgyi Frantsevitch Gause
took a cut at answering Elton's question. Gause's Spitsbergen was a test tube
containing competing species of hungry microbes. In a series of experiments
set down in his landmark text, *The Struggle for Existence*, he fed his captive microbes a broth of bacteria and scrutinized their lethal contests.

Gause's more famous experiments involved two kinds of *Paramecium*, one
superior competitor invariably eating the other's lunch to the loser's ultimate demise. In a less celebrated set of tests, Gause turned his attention "to
an entirely new group of phenomena of the struggle for existence, that of
one species being directly devoured by another." This time Gause pitted
predator and prey in the same tube, caging a harmless, bacteria-sucking *Paramecium* against a relentless protozoan predator called *Didinium*. An insatiable
little monster shaped like a bloated tick, *Didinium* wielded a poisonous dart
for a nose, firing paralyzing toxins into any *Paramecium* it bumped into. Thus
captured, the prey was then devoured whole.

The first meetings of the two were predictably brutish and short, the sequence proceeding as such: Peace-loving *Paramecium*, with no place to run nor hide, gets quickly devoured by the predator *Didinium*. Gorging to its heart's content, *Didinium* soon finds itself alone and hungry, and perishes.

Then in steps Gause, playing God, to level the odds. He adds some sediment to the bottom of the test tube—a refuge, a place for *Paramecium* to hide. *Didinium*, however, knows no other strategy. Seeking and destroying every last *Paramecium* it finds, the savage microbe again eats its way to oblivion. But this time a few lucky *Paramecium* have remained hidden. With the coast clear, they emerge, and soon the world is crazy again with *Paramecium*.

Gause adds one more twist. Every couple of days, he adds a new *Didinium* to the mix. An immigrant. And with that, the little glass microcosm begins producing beautiful numbers. Logged on a line chart, the populations begin tracing sinuous, oscillating waves, prey leading predator through rise and fall, rise and fall, the eternal waves of a predator-prey equilibrium.

It was a beautifully naked, if admittedly clinical, demonstration of the finely and tenuously balanced skills of predator and prey, teetering so delicately on environmental fulcrums. But inevitably, it was the prey in charge, *Paramecium* leading *Didinium* around by its deadly nose.

The world according to Gause was a competitive place. And it was governed from the bottom up. The sun shone, the plants grew, animals ate the plants, other animals ate the plant-eaters, one trophic level to the next, all the way to the tip of Elton's pyramid. The world was in a steady-state equilibrium. It all made perfect sense.

Until a phenomenon called HSS came along.

Hairston, Smith, Slobodkin, and Heresy

In 1960, three eminent scientists from the zoology department of the University of Michigan—Nelson G. Hairston, Frederick E. Smith, and Lawrence B. Slobodkin—wrote a soon-to-be-infamous paper called "Community Structure, Population Control, and Competition," a five-page theoretical rumination published in the *American Naturalist*. The paper was cited and debated so heavily that its authors were thereafter known more simply as HSS. Their proposal earned a nickname of its own: The green world hypothesis.

HSS reasoned that the terrestrial world is green—meaning that it is largely covered in plants—because herbivores are kept from eating it all. And what kept those herbivores from turning the green world to dust, suggested HSS, were predators.

The green world of HSS presented a decidedly different take on how nature worked. The venerable bottom-up perspective had Elton's pyramid progressing smoothly and stepwise from bottom to top, every higher layer inexorably dictated by the lower. HSS, with their hypothetical predators exerting such influence from the top, were fiddling with that sacred pyramid, adding great weight to its peak. In defense of their hypothesis, they cited commonly known plagues of rodents and the outbreaks of insects, apparently following the destruction of their predators. They also raised the legend of the Kaibab.

The Kaibab legend was the classic and controversial tale of predators having the final word. In the 1920s, on the Kaibab Plateau north of the Grand Canyon, the deer population, by all accounts, exploded. And then, the story goes, as the last edible twig was browsed, the population predictably crashed. There was mass starvation, and there were people there to witness it. There was outrage and blame.

The standard explanation, one that has since had a long and potholed ride in the ecology textbooks, pinned the deer's irruption on government trappers, who had cleared the plateau of its wolves and mountain lions. The herd's ultimate demise was therefore triggered by a lack of predators.

HSS used the Kaibab to bolster their case that competition, by itself, fell short of answering ecology's seminal questions on the limits of populations. The phenomenon of predation was now a factor to be heeded. It was a bold hypothesis, abundantly praised and reviled. But what it most fundamentally lacked—as everyone including Hairston, Smith, and Slobodkin would admit—was proof.

At the time the green world hypothesis was being published, Fred Smith, the first S in HSS, had recently taken under his wing a new doctoral candidate. The student was interested in mollusks. More to the point, he—a young scientist after Elton's heart—was a field man fascinated with the workings of food webs. He had situated his dissertation research on a sandy spit bordering Alligator Harbor on the upper Gulf shore of Florida. It was to be a detailed analysis of everything you could possibly want to know about a little clamlike creature known as a brachiopod. But something else had hijacked his attention. There in the clear shallows crawled a varied cast of predatory snails and whelks and conchs, tilling through the sands in an intricate contest of seek-and-destroy. It was, as Robert T. Paine fondly remembered, "a wonderful set of predator-prey interactions."

Museum of Dead Ends

The budding marine ecologist chasing mollusks through the flats of Alligator Harbor had had a bumpy transition from his preordained destiny as an ornithologist. Paine had grown up in Cambridge, Massachusetts, a bird-watching prodigy in one of America's hotbeds of bird-watchers. Paine was the youngest member ever of the exclusive Nuttall Ornithological Club, a who's who in American bird-watching. By the age of fourteen, Paine had keys to Harvard University's Museum of Comparative Zoology, where he was free to peruse one of the country's most precious collections of bird specimens. It was all very interesting, but Paine ultimately longed to be back outside, to study them alive.

By the time Paine entered Harvard as a college student, he—with a sense that bird-watching might not be the most practical profession—had gravitated to the geology department, to pursue his second love of paleontology. Under the aura of such legendary names as Whittington and Romer, he studied fossil shells from sedimentary rock laid down in the Silurian seas four hundred million years ago. The rocks were studded with ancient marine animals. Paine looked hard at crinoids, feathery-limbed cousins of the starfish, bearing broken and missing appendages. He was intrigued by the spiny creatures that came obviously dressed for defense. Paine was imagining them all alive and chasing each other.

In his senior year, Paine was lured from his ancient seas and detoured back to the bird world, under the fatherly wing of one Ernst Mayr. One of the great figures in evolutionary biology, Mayr was also a world-class ornithologist, who set about shaping his twenty-year-old protégé in the traditional vein. He sent Paine through the library stacks and museum collections, had him staring at series of bird skins, deciphering subtle difference in feather patterns. Thirty-five, forty hours a week—"Mayr didn't stop to consider whether I might have any other interests like sports or girls," as Paine likes to tell. But Mayr's was an offer—coming as it did from the godfather of evolutionary biology—that no self-respecting Harvard undergrad could refuse.

Mayr temporarily freed Paine to accompany a Harvard ornithological expedition to the tropical forests of southern Mexico. In this strange and fearsome place crawling with jaguars and fer-de-lance vipers, Paine's job was to collect birds, which he did with the museum's tool of choice. For eight hours a day, Paine plugged away with his shotgun, bagging specimens.

When he returned, Mayr was waiting with questions. They had to do with distinguishing variations in plumage and bill length and eye color. Paine answered with questions of his own. What he rather wanted to know was what had lain inside their stomachs. He wanted to know how they lived, how they all interacted, how the system worked.

"Those are interesting questions," said Mayr, looking Paine in the eye. "And you'll never make a professional ornithologist asking those questions."

"It was 1954," remembered Paine. "And the world wasn't ready for such thinking."

After a two-year stint in the army, Paine was back on the doorsteps of academia, applying for a doctoral program at the University of Michigan. Shaking off his bad date with ornithology, he vowed again to become a paleontologist after all.

Paine's vow died a quick death. Michigan's curriculum offered "a thousand and one courses in ichthyology, paleontology—all totally boring," thought Paine. On the last day of an exhaustive admission process, the man who would become his adviser saw the glaze in Paine's eyes and stepped in, telling him: "Why don't you change fields? You go get your admission material for the department of zoology and I'll get you accepted. You don't want to be a geologist. You want to be an ecologist."

"You're right," answered Paine.

Gastropod Warfare at Alligator Harbor

No more dead stuff for Robert T. Paine. Remembering his Harvard days contemplating interactions of four-hundred-million-year-old fossil shells, Paine soon found himself standing shin deep in Alligator Harbor, watching the live versions in active battle. There were eight of them: *Pleuroploca gigantea, Fasciolaria tulipa, Fasciolaria hunteria, Busycon contrarium, Busycon spiratum, Murex florifer, Polinices duplicatus,* and *Sinum perspectivum*—otherwise known to any amateur collector by such names as Florida horse conch, tulip snail, lightning whelk, shark's eye moon snail. But these were something other than a benign collection of shiny shells in a museum display case. In life, all were working predators. And all were earnestly involved in eating each other.

Inside these crawling ornaments beat the hearts of pure predators, wielding single-minded intents and one of the meaner pieces of weaponry in the animal kingdom. All slugs and snails of the molluscan group known as the gastropods come with a muscular strap of tooth-studded flesh called the radula. It is a tool in various forms analogous to the workman's rasp, file, drill, and chain saw, the critical distinction being that the radula is most often employed to destroy and consume living flesh. Among the carnivorous clan of gastropods, the radula is commonly wielded against its cousin snails and clams. Plowing through the sandy bottoms, the hunting snail bumps against another shell, and the struggle begins. If the defender comes with an open-ended shell, such as another snail, the process is straightforward and quick. The attacker's proboscis snakes its way inside and the eating commences, rasping away at the hapless homeowner cornered in its own fortress. If the prey is a clam or mussel, valves tightly closed, the radula emerges to begin the slow work of drilling through the shell. It may take two days to penetrate, but the snail has patience, working the drill slowly from this angle and that, like a diamond bit through rock. Once the shell is pierced, the proboscis pushes inside, and the radula goes to work on soft flesh, shredding the clam where it lives. What remains when the snail leaves is an empty clamshell with a clean circular hole, as if pierced by a bullet.

Some snails dispense with the precise drilling in favor of brute force, prying their proboscis through an unguarded opening. If the opening is too small, they chip away at it with a sharp edge of their own shell until the proboscis fits. And then again the radula finishes the job.

This was the brand of carnage Paine had come to witness. Inside this outwardly meek world of the mollusk, Paine was spectator to a vicious society of pursuit and piercing wounds, gang-tackling, winner-take-all contests of strength, breaking and entering, piracy, and cannibalism. Paine would walk these sandy battle shores and methodically tally the casualties, marked by drill holes and chipped shells, sometimes catching predation in the act. In the end he would note the winners and losers, who were eating whom. In the end, big uncommon shells were eating small and numerous shells—a pyramid of predators.

It was all so very Eltonian.

For better or worse. These, like Elton's seabird observations, were just that. They were not experiments; they were observations. They showed only that big shells ate smaller shells. There was no accounting for how the individual acts translated to numbers lower on the food chain. They proved nothing about why the world was green. HSS was still just a balloon of a theory waiting to be either bronzed or popped by experimental evidence.

In order to truly test how much the top predators held sway, one would have to do more than watch; one would have to dive in and start moving pieces around, to manipulate the predators, to measure the before's and after's, the with's and without's. For starters, one would need a system a bit more tractable than the confounding gang of little predators trolling Alligator Harbor.

"The Dumbest of All Experiments"

In 1961, Paine defended his dissertation to a committee including Nelson Hairston and Fred Smith, the first two thirds of the infamous HSS. Finishing up at Michigan, he again headed for the coast, this time to the Pacific sands for a stint of postdoc research at the Scripps Institute of Oceanography in La Jolla. There he studied a sea slug that ate other sea slugs—"Just sucked 'em in! An absolutely violent act." It was molluscan mayhem all over again. It seemed that everything was coming up predators for Paine.

In his free time, Paine would venture down to the Scripps wharf, where he found clinging to the pier pilings massive beds of California mussels. And they in turn were under attack, by leagues of ochre starfish. Paine dwelled upon the scene.

Mytilus californianus, the California mussel, is a blue-gray bivalve, cousin to the clam, with the competitive advantage of a secreted fiber cable called the byssal thread, by which to anchor itself on hard surfaces. The byssal thread is one large reason for the extensive range of the inadequately named *Mytilus californianus*, a creature of surf-exposed rocks and pier pilings from the shores of Alaska to Baja.

Mytilus eats by opening its shell and straining plankton from seawater. Snuggled tight against its neighbor, lazily yawning to its belly's delight, apparently commanding the life of Riley, *Mytilus* does occasionally suffer enemies. Shorebirds and crabs will take advantage of a mussel in a vulnerable moment. Big whelks will sometimes drill their way through *Mytilus*'s shell. These it takes in stride. But if a mussel could feel fear, *Mytilus* would be positively quaking in its shell at the approach of one *Pisaster ochraceous*.

Pisaster, less formally known as the ochre starfish, is big and brawny, its arm span reaching eighteen inches in some of the giants of the race. It is a hardy, shallow-water species, capable of withstanding strong ocean surges, big swings in temperature, and desiccation when left stranded by the tide. Like *Mytilus*, *Pisaster* has mastered life over the rocky breadth of the North American Pacific coast. Unlike *Mytilus*, *Pisaster* is pure carnivore, and mollusks are its favorite meat.

With high tide, *Pisaster* comes hunting, creeping out of the depths, up the pilings, up the rocks, to where the mussels have huddled. Gliding about on thousands of tiny hydraulic feet, *Pisaster* gives chase at speeds better appreciated with time-lapse photography, yet infinitely faster than any rock-bound mussel. Upon bumping into the prospect of prey—the starfish being essentially blind—*Pisaster* crawls on top, tiny tentacled feet sniffing for the scent of food. Edibility thus determined, the starfish centers itself over the mussel and stands it up, hinge-side down. The mussel is now positioned so that if it were to sneak a peek, it would be staring straight into the mouth of Pisaster. The starfish embraces the mussel, envelopes it like a tent. Its arms clamp around the opposing shells and begin prying.

The pull is not explosive, but steady and unyielding. It is the ratcheting squeeze of a python in reverse. It is less a matter of grip strength than endurance. Mytilus's muscles begin to tire. A tiny gap opens between valves. The fight is lost.

Pisaster exploits the opening, injects a dose of stomach juice into the gap. The weakening mussel is being digested where it lives. The opening widens, and in comes Pisaster with its entire stomach. Extruding its gut out its mouth and into the home of Mytilus, Pisaster begins feeding externally. Two days later, depending on portion size, the starfish leaves a vacant shell and crawls on to another.

The preceding explanation implies a Pisaster free-for-all. But in practice, there is a bit more cat-and-mouse to the contest. A starfish prowling much beyond the low tide line—up where the masses of mussels lie huddled—is an exposed and vulnerable starfish. Crashing waves can batter and dislodge it. The attacking starfish advances at its increasing peril. Too long out of water, its hydraulics begin to falter, and a strong and stubborn mussel gradually gains the upper hand. Both starfish and mussel are vitally tethered to the sea by invisible chains, but it is the mussel that is able to endure a critical step higher up the tide line.

On the Scripps wharf, Paine stood before this cobbled pavement of orange starfish humped over doomed mussels—Pisaster chewing on Mytilus—and experienced, in his words, "Love at first sight."

He pried off a starfish to ponder the mussel smothered in its lethal embrace. It struck Paine that he could intervene in the predatory act as easily as walking up and plucking a starfish. He began to sketch an experiment, as one might a theatrical play. He already had in hand the Pisaster-Mytilus dynamic, his lead predator and prey. He penciled in a few clams, some crawling gastropods, a few worms to play supporting roles in his ecosystem. Now he would measure the predator's power in the most logical way—he would remove it. He would remove *Pisaster* from an intertidal community and see how everyone else in the food web fared.

"It was clearly the dumbest of all experiments, removing starfish to reveal some idea of how their prey would respond," said Paine. "It just struck me as bizarre that nobody had thought to remove *Pisaster* to see how *Mytilus* would respond. To see how the system worked. What roles the species played, rather than just their addresses."

This was territory that even his hero Elton hadn't quite explored. As rock-solid as Elton might have been in his reasoning—who would argue that the greenery beneath the seabird cliffs of Spitsbergen had not been fertilized by fish-eating birds?—without systematic experimentation, his observations still amounted to conjecture.

Paine's sketch grew into a full-blown proposal. He would find his model community and test its vital structure. And away he went, with his first professorship awaiting him at the University of Washington in Seattle and a prestigious grant from the National Science Foundation to carry out this important new investigation on the role of predation in the web of life.

There was only one small catch. The only place Paine knew for sure to find his concocted ecosystem was still on paper.

Playing God in a Tide Pool

Paine arrived at the University of Washington in 1962 and went to stake out his study plots. He searched the shores of the San Juan Islands, to the far ends of Puget Sound. No luck. He knew his creatures existed, he knew from the literature they indeed lived together. He just couldn't find them.

As it happened, part of Paine's teaching assignment included a field trip to the coast, a four-hour drive west from Seattle to the tip of the Olympic Peninsula. In May of 1963, he led his students to a remote beach of rock and sand for their introduction to the natural history of marine invertebrates. And there on the rocky shore of Mukkaw Bay, he saw it laid out in front of him, waiting. My god, here it is right in front of my eyes, thought Paine. "It took me half a millisecond to say, 'Wow, this is what I've been looking for.' " A month later, Paine was back, chucking starfish, embarking on one of the classic studies in community ecology.

Twice a month, Paine would travel from campus in Seattle, board a ferry across Puget Sound, and drive the long coastal road heading west, tracking the Strait of Juan de Fuca toward its meeting with the Pacific. Approaching the tip of the Olympic Peninsula, he would cross into the Makah Nation, where earlier in the century Makah tribesmen hunted gray whales with harpoons thrown from canoes dug out of cedar trees. Then he would make a short inland jaunt across the peninsula, to the coast of the Pacific and the little cove of Mukkaw Bay.

Mukkaw Bay forms an arc of dark sand and rock, where the face of the continent is still being freshly ground. At the northern tip of the arc, the Pacific Ocean and what used to be headland of the Olympic Mountains are still at odds. There standing in the water is a sea stack, quintessential landmark of the Northwest coast, a naked pillar of stone still defying the surf. And lying beside it is a rubbled tongue of rock, still struggling to keep its nose above water with every incoming tide. Here, according to the scriptures of ecology, lay holy ground.

There were no beach stands, no umbrellas, and for the most part, no people. Paine would walk to the outcrop, a labyrinth of rock walls and channels surging with water on the incoming waves—a flooded badlands. It was a place of beauty, of pools glimmering with multicolor gardens of red and brown kelp and green anemones, pink coralline algae and purple sea urchins, yellow tunicates and blood stars.

The rock was pitted and grooved, the seascape constantly changing with the pendulum swing of the tides and thrashing storms. It was this ironic harshness—these rocks to hold on to; these shrouds of fog and pools of wetness providing refuge against the desiccating exposure of low tide; this turbulent influx and flush of marine nutrients—that gave the rocky shore its vitality.

It was a good place to cling, providing you came with the suction feet of a starfish, or the anchoring cables of a mussel. It was something more the treacherous place if you were a tall, two-legged creature evolved upon

African savannas. Over this terrain the six-foot, six-inch Paine invariably came weaving and bounding, twice a month in spring and summer, once in winter. He set up his study on two adjacent twenty-five-by-six-foot sections of rock, featuring a ledge to stand on and a handily accessible wall of marine animals exposed by the tide. Here he had all the ingredients for deciphering what a community ecologist would formally characterize as a Pacific Northwest intertidal food web.

The rock was a stratigraphy of marine life. Near the top clung scattered clumps of little acorn barnacles, they in turn interspersed with clusters of the big and beautiful goose barnacle, intricately cloaked in harlequin shells. Below the barnacles formed a blue-gray band dominated by the California mussel, packed ear to ear.

In the tidal pools below, the neighborhood suddenly flowered into a marine melting pot—two species of Chinese hats called limpets, four species of variously colored algae, a big green anemone, a sponge, and a sponge-eating slug called a nudibranch. And separating one from the other, oddly contrasting midway up the rock, was a constellation of big orange and purple starfish, *Pisaster ochraceous*.

The pecking order was well established, and starkly simple. *Pisaster* fed on all shelly creatures of the rock; none fed on *Pisaster*. In this kitchen-size ecosystem, *Pisaster* was the undisputed king—which Paine was about to dethrone.

Each month with the low tide, Paine would, as he put it, "waltz on down to the sacred ground," count the number of starfish on the experimental side, pry them off, and heave them as far as he could into the surf.

"From then on," said Paine, "it was an arms race between me and *Pisaster*."

It was not always clear who was winning. Every month, *Pisaster* would march back from the depths and reappear on the rock, terrorizing mussels and barnacles like a pack of wild dogs after cats up a tree. And Paine would heave, and heave again. There was a lot of starfish, and a lot of heaving. Paine's pitching arm suffered rotator cuff injuries. After a couple of days' battle with the rough skin of *Pisaster*, the skin on Paine's hands would begin to peel.

The longer Paine persevered to throw *Pisaster* from the rock, the more the rock's assemblage changed. It began with the acorn barnacles spreading beyond their old bounds. Room for other species soon fell short. Chitons and limpets and some of the algae species decamped. By September the barnacles had covered three quarters of the available space. By the following June, it was the barnacles' turn to be evicted.

Without *Pisaster* to police the grounds, *Mytilus* was now seriously flexing its muscles. It emerged as the fiercest competitor on the block. It swarmed down, staking out lots with its byssal threads. The rock grew black under a spreading pavement of mussel shells. Paine took inventory. Seven of the fifteen species at the start of the experiment were gone, and all the rest but *Mytilus* were fast heading for the exits.

Paine could plainly see where things were heading. After just a year, he had enough on record to share with the scientific world. He wrote it up, his resulting paper appearing in the January–February 1966 issue of the *American Naturalist*, under the title "Food Web Complexity and Species Diversity." His hypothesis was stated with this sentence: "Local species diversity is directly related to the efficiency with which predators prevent the monopolization of the major environmental requisites by one species."

In other words, the suggestions of Elton, the test tubes of Gause, the theorizations of HSS, now had the cachet of experimental backing in the field. Species indeed were highly interconnected by a web of interactions, a web made more stable by a complexity of species, a web over which predators could wield inordinate powers.

Paine's own interpretation was at turns dutifully restrained and subtly intrepid. He noted both the inherent limitations and sweeping possibilities of this tiny point of rock on an obscure outpost of Pacific shore. "This hypothesis offered herein applies to local diversity patterns of rocky intertidal marine organisms," he wrote, "though it conceivably has wider applications."

Those wider applications would turn Robert T. Paine's "dumbest" little experiment into one of the most cited papers in the history of ecology. Solid in its simplicity, Paine's heroic portrayal of *Pisaster* had laid siege to the stigma that ecological webs were beyond the realm of experimental science. And if one took the serious effort to look, they could very well find predators pulling critical strings in that web.

In a follow-up paper for the *American Naturalist*, unassumingly titled "A Note on Trophic Complexity and Community Stability," Paine coined a term for species of *Pisaster*'s importance—species whose impact far overcompensated for their relative scarcity. Such species were hereby "keystone species," wrote Paine. Remove them from the archway of life, and the whole structure comes crashing down. With that, community ecology added a powerful new concept to its lexicon, embodied in Paine's shining new star.

Paine and *Pisaster* spawned a world of new research questions. Which predators matter, which ones do not? What about the deep sea, or for that matter the terrestrial realm? Any keystone predators out there? That is, just how common was such a predator of *Pisaster*'s impact?

Somewhere along this line of inquiry came a more anxious slant on the question. A new generation of ecologists freshly weaned on the weekly wonders of *Wild Kingdom* and the ominous horizons of Rachel Carson's *Silent Spring*—presensitized as they were to the specter of mass extinction and the cause of conservation—were reading Paine's results with a more concerned eye. They were wondering about such predators as wolves and killer whales, great cats and great white sharks. *Pisaster* had proved that certain predators, by their mere presence, could bolster the diversity of life. But just as easily, once removed, that benevolent hand could be replaced by a phantom fist, knocking species off the planetary rock, as it were, overhauling the living landscape to simpler, cruder states. To these ecologists, the monoculture of mussels crowding Paine's predator-free rock was handwriting on the wall. It raised the larger question, How *are* all those big, fearsome, topmost predators doing, anyway?

TWO　Planet Predator

No one seems to approve of predation but, like sin, it is not often that anyone suc-
ceeds in stopping it for an appreciable length of time.

—**Durward Allen**

Genesis: The First Bite

Not long after a swirling cloud of cosmic dust and gas condensed to become the solar system, life arose on the third planet from the sun. Earth in
those times, some four billion years ago, was an alien realm of rock and water under violent assault, showered in lava, shocked by lightning and bom-
barded by ultraviolet rays beating unhindered through an atmosphere
starved of oxygen. Simple organic molecules, synthesized with the flashing
of heat, spiced the oceans, eventually bonding to produce a protein-coated
cell encircling a spiraling, replicating strand of nucleotides nicknamed DNA.
For a billion or more years, life was just such a simple, single-celled affair,
of bacterial microbes feeding off the sun, reproducing without sex. Then
one day, as the theory goes, one of those cells somehow swallowed its
neighbor. And life, as they say, would never again be the same.

It was a natural step or two beyond that for predator and prey to add
spermlike tails and fluttering cilia of rudimentary self-propulsion, the first
glimpses of the chase. With this new race of supercells now roaming the
seas, evading and pursuing and eating each other, size too became an issue.
For the sake of survival, it was a good idea to be bigger than your attacker

could swallow. Two ways of doing that were either to grow a bigger cell or to join with other cells into a more forbidding collective mass. An answer to increasingly dangerous times was under development, in the form of multicelled organisms.

By introducing "hazard to complacency"—in the words of paleontologist Stephan Bengston—predation had begun transforming a generic world community of microbes to a biosphere humming with complexity and gigantic potential. For a long time that potential lay smoldering among the miniature masses, until one more defining moment in the history of life, when life itself erupted.

The Big Bang

Marine fossils paint an idyllic scene of animal life in its infancy, off the naked shores of the Precambrian continents some 670 million years ago. Coralline fronds arch lazily from the ocean floor, ancestral jellyfish undulate in the currents, marine worms plow through the ooze.

Rolling forward a quick one hundred million years, to the dawn of the Cambrian period, the seascape abruptly changes. Animals suddenly appear cloaked in scales and spines, tubes and shells. Seemingly out of nowhere, in bewildering abundance and variety, the animal skeleton emerges.

Paleontologists have long wrinkled their brows trying to explain why— after at least one hundred million years of soft, serene, multicellular existence—life so hurriedly turned hard. The startling abruptness of it all had troubled Charles Darwin to the point of doubting his own theory of evolution. "To the question why we do not find rich fossiliferous deposits belonging to periods prior to the Cambrian system, I can give no satisfactory answer," he confessed in his masterwork, The Origin of Species. "The case at present must remain inexplicable: and may be truly urged as a valid argument against the views here entertained."

Sophisticated hypotheses have been volleying back and forth for decades, some linking the skeletal genesis to changing chemistries of the seas and skies. But wounds discovered in the fossils themselves suggest the sudden hardening of life was more than just a chemical reaction. It was an arms race.

The Cambrian explosion, beginning some 565 million years ago, created in a 10-million-year spree of evolution the blueprint for virtually every major body plan of animal life existing today. It cast the die, in variously deceptive guises, of everything animal—shrimp to giant squid, ant to elephant. And as much as that revolution likely had something to do with rising global levels of atmospheric oxygen (thus enabling turbocharged metabolisms and leaps in body size), it by no small coincidence also came with an evolutionary flourish of lethal weaponry and armored, mobile defense.

The Cambrian was a time for the meek to start watching their backsides. Among the predatory guild was Sidneyia, helmet-headed like a horseshoe crab, with two columns of grasping limbs. The sly Ottoia was a chunky burrowing worm ambushing little shelly creatures from below, vacuuming them through its muscular proboscis—the feat confirmed by whole little fossil mollusks found inside Ottoia's fossil intestines. By modern standards the Cambrian collection pushes the bounds of credibility. There was Opabinia, anticipating B-grade Martian fauna by half a billion years. Opabinia was an elongated cockroach with five knobby eyes and a serpentine proboscis. The hose nose ended in a mighty claw that evidently cherry-picked careless creatures from the bottom of the sea and delivered them to the mouth. When the paleontologist Harry Whittington of Cambridge University first shared his reconstruction of Opabinia to a gathering of serious-minded colleagues, they broke out laughing.

Among the prey swam the soft-bodied Pikaia, a graceful, undulating ribbon of flesh, unglamorously hunting detritus. In navigating the Cambrian war zone, Pikaia obviously relied on something other than brawn. The relevance of Pikaia's survival, mentioned here mainly for anthropocentric purposes, stems from its unique wiring. The little worm was internally governed by a single rod of nerves running head to tail, the founding hallmark of the phylum Chordata. The descendants of Pikaia would eventually sprout the fins of fish and fill the oceans, the meaty thighs of tyrannosaurs to stomp across the land. Some would spread the wings of birds and cruise the continents by air. Others would come to wear trousers and classify fossils.

But back to the true heroes of the battleground. Among the Cambrian prey crawled the trilobites, the lovable cockroaches of the sea, hardened and armored to the hilt. The dreamlike *Hallucigenia*, named for its catalogers' first impressions, was a multilegged oddity like a centipede, but bristling with spines. There also crept *Wiwaxia*, a sluglike beast sheathed in a chain-mail suit of armor and—just to make sure—a minefield of dagger blades springing from its back: A headless armadillo crossed with a tailless porcupine.

Some of the fossils that show what these 540-million-year-old armored animals looked like also suggest what they were steeling themselves against. Some of the spines of *Wiwaxia* appear to have broken and healed. Holes in the armor of certain trilobites suggest they had been bitten.

Bitten by what, nobody could at first imagine. Because the terror of the trilobites was, for nearly a hundred years, still lying in little, innocent-looking pieces. In 1886, Canadian paleontologist J. F. Whiteaves uncovered a fossil from a British Columbia shale quarry, of a spiky creature about three or four inches long—the tail end of something shrimplike, guessed Whiteaves, who, thus stumped, named his creature *Anomalocaris*, or "odd shrimp." Another paleontologist, Charles Doolittle Walcott, in a nearby outcrop called the Burgess Shale, later found a circular creature, radially divided into thirty-two segments, pineapple-wise. Walcott named it *Petoyia* and assumed it a jellyfish.

For decades the two creatures appeared in various artists' reconstructions of the wondrous Cambrian seascape, each going about its own business. It wasn't until 1985 that Harry Whittington and his colleague Derek Briggs, upon reexamining the specimens (by then lying dormant seventy years in the bowels of the Smithsonian Institution), decided the two creatures were actually body parts of one another. That four-inch shrimp of Whiteaves's turned out to be one of *Anomalacaris*'s claws. That jellyfish of Walcott's was its mouth. Once pieced together those claws and mouth translated to a monster that in its time approached four feet long, as huge and deadly as anything that had ever moved on Earth. As Briggs and Whittington proudly pointed out, *Anomalocaris* was the *Tyrannosaurus rex* of its time.

Anomalocaris glided through the seas on undulating wings like those of a stingray, snatching prey with its hooked claws and crunching them in that deadly pineapple slice of constricting blades. When Briggs and Whittington simulated the bite of *Anomalocaris*'s circular guillotine, it formed a W, matching suspiciously the outline of wounds not uncommonly found in trilobites among various Cambrian fossil beds. The wounds suggested more than a demonstration of shearing force. Those disfigured trilobites, by their very presence, signified creatures that had gotten away. Their battered armor represented not failure, but defenses that had served their purpose. Given that the trilobites were to swim the Paleozoic seas for another three hundred million years, they certainly qualified—by any survival-of-the-fittest criterion—as evolutionary successes.

Which in turn put the monster *Anomalocaris* in a more benevolent light. If that which didn't kill the trilobite indeed made it stronger, then *Anomalocaris* must have been one formidable strength trainer. The Cambrian was a time in which "ways of eating . . . and ways to keep from being eaten proliferated," wrote Briggs and Whittington. And *Anomalocaris* had pushed the limits on both ends.

The Cambrian frenzy was the first unmistakable appearance of the modern complex food web, woven not only with producers and consumers but also consumers of consumers—those superpredators capable of eating all others. It served as the first hard evidence for predation as a major architect of life's myriad form and function.

Beauty in a World of Wounds

To what degree, the impact of predation, remains the eternal question. It is the one that Geerat Vermeij, an evolutionary biologist who holds an office at the University of California at Davis, has spent most of his life asking.

When Vermeij was nine years old in East Dover elementary school in New Jersey, his fourth-grade teacher brought to class a collection of shells from Florida. Vermeij examined the shells like no other student ever had. He turned them over in his hands, as if manipulating a puzzle, his fingers gliding over undulations, bumps, and ridges, noting the sharpness of edges, the textured exteriors and glassy interiors. There was something altogether different and more exciting about these warm-water shells than the coarse and chalky cockles and razor clams he'd collected as a toddler on the shores of his Netherlands homeland. "How could one explain a shell as odd as the lightning whelk, with a spiral crown of knobs at one end and a drawn-out spout at the other?" he later reminisced. "Why was its interior so stunningly sculptured with smooth, evenly spaced ribs that spiraled away beyond the reach of my fingers?"

Vermeij had never seen such beauty, which had nothing to do with the fact that he'd been totally blind since the age of four. (Born with a rare form of glaucoma, Vermeij had seen only dim shapes and colors before he lost his eyes to the disease.) With his practiced fingers, Vermeij viewed the shells in three-dimensional clarity. Their strength and complexity contrasted so conspicuously with lesser-endowed shells from the cold waters of the North Atlantic. "Why should this be so?"

Immediately thereafter, cigar boxes began filling with the new collection of young Gary Vermeij, self-confirmed conchologist for life. After blazing his way through high school (top in his class), then Princeton (taking Braille notes with a stylus, hiring readers to convey the printed literature), through his graduate careers at the University of Maryland and Yale, Vermeij set about gathering the shells of mollusks from the oceans of the world, Costa Rica to New Zealand, Micronesia to Madagascar.

From his global excursions in search of shells, Vermeij began to develop a curious picture of molluscan architecture. "How odd that all the cold-water shells from New Zealand are so thin," he thought.

Vermeij would hunt his mollusks by wading slowly through the shallows, toes probing the sand and seaweed for shells, or crawling hand-and-knee upon the shore, acquainting himself with the topography, listening for the rumble of an oncoming wave, the echo of an approaching crevasse. Lifting a specimen, he would run his fingers over the contours, seeing by feeling the lip and protuberances, the aperture, the spiral formations, the ridges and valleys, over a living landscape of calcium carbonate.

The tactile explorations came with their hazards. Vermeij in his wanderings would be stung by a stingray in Panama that as he recalled, was "flapping as it sought in vain to free its barbed tail from my foot." He would be bitten by a moray eel in Polynesia. In Guam, Vermeij mistakenly caressed a live stonefish, whose dorsal fin is a venomous hypodermic needle capable of inflicting one of life's most excruciating pains, and occasionally the lesser agony of death.

As Vermeij collected, the artist in him grew to despise the marred and broken shells that so often passed through his fingers. He would toss them aside as one would a chipped plate from a set of fine china. While Vermeij was still in his teens, his brother gifted him with a special shell painstakingly procured from the Philippines. A gem to the untrained eye, under the scrutiny of Gary Vermeij's fingers it might as well have been a lump of cubic zirconium. "It was with a sinking feeling that I realized the specimen was badly flawed. Running in a jagged line from one end of the shell to the other was a scar, a rough, ugly welt rudely interrupting the otherwise smooth contour of the shells' pleasing conical shape." To Vermeij, the scar was "a blemish, an insult that some careless human collector must have inflicted on the specimen."

It seemed the seas were full of such disappointments. In 1970, while wading tropical tide pools in Guam, Vermeij's colleague handed him a cowrie, a beautiful piece of polished shell, otherwise ruined by a hole in what had been the animal's roof. Vermeij sighed and chalked it up to another "crime against nature"—probably the work of waves pounding upon the shallow reef. When his colleague mentioned he'd seen shells broken in this very way by crabs he kept in his aquarium, Vermeij shrugged, filed the thought, and forgot about it.

It wasn't until two years later, pondering yet another summer of broken shells, that the flawed gift from the Philippines, the wave-battered cowrie from Guam, the crab in the aquarium, flashed to a single coalescing revelation in Vermeij's mind: those disfiguring scars and welts, those elaborate spines and thick buttresses, those tight coilings and tiny openings—as if the occupants had spiked their house, locked the doors, and huddled in the corner—were neither fashion statements nor the fallout of dumb luck. These were the armaments and injuries of battle.

The persnickety collector therewith swung about-face, eschewing the most elegant of specimens for the roughest of characters, gathering battered old battleships and crippled goners, seeking the wrecked and wounded from around the globe. "Suddenly, all those ugly broken shells that I had previously dismissed as unacceptable specimens became mines of information," said Vermeij. "Once I recognized the evolutionary importance of scarred shells, I encountered repaired damage everywhere I looked."

Vermeij found destruction everywhere, but curiously more so in the heavily armored shells of the tropical Pacific than in the lighter fortified shells of the Atlantic. The reason, he first ventured to say in a 1974 issue of the journal *Evolution*, amounted to the more varied and lethal cast of predators cruising the Pacific sea bottom. The variation in shells was evolution as an arms race.

Back again in Guam, Vermeij resumed exploring his new view of evolution with a small experiment easily appreciated by any little child who has ever, in a moment of sadistic mischief, thrown a grasshopper into a jar of raging ants. Into an aquarium of variously fortified snails, Vermeij introduced two of the reef's most powerful crabs. *Carpilius maculatus*, a bulky brute measuring nearly five inches across the back, wielded massive, stubby claws studded with blunt crushing teeth; *Eriphia sebana* was a smaller but comparably serious crab, with longer and unmistakably lethal claws. The dangerous crustaceans went into the tanks with the defensive snails. And together, Vermeij and his wife, Edith, stood back and listened, waiting for the inevitable explosions of shell.

The crabs would first tend to focus their attack on the rim of a snail's aperture. Wrestling for position, a pincer would find the point of maximum leverage.

Pow!

Small thin snail meets big powerful crab—an obvious mismatch.

Kapow!

Not to imply that the battles were one-sided bloodbaths. Sometimes the struggles went an hour. Often enough, powerful pincers met impregnable shells. Many a stubborn mollusk sent an exhausted crab to defeat, the mollusk's shell battered but intact, still housing one whole and securely ensconced snail. At times it seemed hard to decide which was the underdog.

Among the most heavily armored of snails was a battleship named *Drupa* morum. The knobby, fortified shell wall of Drupa morum ran two to three millimeters thick, the creature inside secured behind a tight, constricted opening that barred probing pincers. Drupa, the living fortress, sent away many comers to defeat.

But then, there came the she-crab named Railroad Tie. The biggest most awesome specimen of Carpilius maculatus in Vermeij's collection was nicknamed for the heavy log needed to bar the lid of her aquarium against her escape. Railroad Tie had never met a snail she couldn't break. In the marquee event of Vermeij's Coliseum, the unstoppable offense of Railroad Tie was pitted against the drum-tight defense of Drupa morum. On meeting the impenetrable mollusk, the big crab with her extraordinary right claw bypassed the customary perimeter assaults on the snail's rim. Her claw enveloped the entire shell of Drupa. She started squeezing with a force of some twelve hundred pounds pressure.

Drupa's shell held.

Railroad Tie squeezed.

Forces collided in straining tension, muscle against mineral.

Pow!

Of course, a few busted snails in an aquarium did not serve as proper foundation for a major contribution to evolutionary theory. Vermeij went on collecting, gathering the pieces, synthesizing them into a predatory worldview of life. The Cambrian explosion had been the first great show of predatory power, he agreed. When he looked 350 million years down the road, to the onset of the Mesozoic era, he recognized yet another.

By this time, there were crabs crawling about the Mesozoic seafloors with the specialized crushing claws that in their day might have invoked the same sort of fear as a modern day Railroad Tie. There were predatory sea stars with clam-prying arms. The waters by this time were swimming with fish, giant bony monsters with massive jaws and teeth. Even Vermeij's adored ones had grown particularly lethal. The mollusks had developed their signature weapon, the radula, that tonguelike strap of flesh studded with steely teeth—the living chain saw.

The "Mesozoic marine revolution," as Vermeij provocatively named it in a 1977 paper, was a time of profound biological reorganization. To his mind it was an era of escalation, of big predators growing bigger and faster, of prey growing swifter and craftier. It had brought Vermeij to appreciate the power of the struggle.

"I think the world is fundamentally about competition, and predation is one very important form of competition," he said. "It's a world of 'I want those resources and I'm going to compete for those resources.' And of course the other side of the coin is 'I want to defend myself.' Everything in biology, really, is an arms race."

Shells were Vermeij's window into life. And life, according to Vermeij, was a beautiful array of art forms, forged in the blood-soaked battleground of competition, ever escalating toward new heights of creative design. And for hundreds of millions of years, by all appearances, that was indeed true.

Escalation

By the time the monster *Anomalocaris* burst from the quiet depths of Precambrian time, evolution had already sent an array of predatory lines fanning outward and growing in leaps. In time the carnivore lifestyle would produce meat-eaters of monstrous, metaphor-straining proportions. By 350 million years ago, there were predacious fish the size of buses: *Dinichthys* was thirty feet long, with a tail like an eel and a bony, jagged-toothed skull suggesting a jack-o'-lantern carved from a wrecking ball. By the golden years of the dinosaurs, not long before their meteoric demise 65 million years ago, reptilian evolution had populated land and sea with dragons. There were forty-foot sea lizards called mosasaurs, and the shores housed a fifty-foot crocodile called *Deinosuchus*, which probably preyed on dinosaurs. Sharks appeared soon after, and in time would produce *Carcharodon megalodon*.

Ancestor of the great white shark, *Carcharodon megalodon* is typically pictured nowadays as a set of wickedly toothed jaws gaping nine feet across, with a group of six dour technicians in lab coats and white shirts comfortably framed inside. The creature once behind those jaws stretched more than forty feet long. If an ox could swim, megalodon could have swallowed it.

Megalodon might well have been the all-time alpha predator of the ocean, but the land, of course, had *Tyrannosaurus rex*. King of the tyrant lizards, star of the big screen, centerpiece of natural history museums—T. rex, the baddest meat-eater that ever walked the land. And also of late, T. rex, the favorite punching bag of paleontologists arguing the validity of that reputation. (Was it really swift enough to chase down that jeep in Spielberg's *Jurassic Park*, or would it more likely have had trouble keeping pace with a small herd of schoolgirls hopscotching down the sidewalk? Was it a high-octane superpredator, or a plodding, parasitic scavenger bullying smaller predators off their kills?)

Either way, there were those undeniable teeth, finely serrated daggers protruding six inches from massively muscled jaws. Curious biomechanics now love to play with those jaws, applying their physics formulae and digital simulators to arrive at biting forces peaking at four tons per square inch.

If T. rex wasn't the one killing all those *Triceratops* and duck-billed hadrosaurs mingling alongside it in the fossil quarries (some with tooth scars traced to T. rex itself), something other than old age and accidents likely was. There had to be some reason beyond decoration to explain the quadruped tank called *Ankylosaurus*—a reptilian beast the size of a rhino, with a knobby, bone-plated shield draping upon its back, and a tail terminating in a medieval mace of bone and tendon. *Ankylosaurus* was the quintessentially defensive dinosaur.

For those of the T. rex-as-scavenger persuasion, paleontologists have offered up a line of smaller, sleeker predators that, in their day, might have acted as giant killers in the ways of modern lion prides and wolf packs. Living alongside the lumbering tyrannosaurs was a related family of small, hyperpredatory dinosaurs called Dromaeosaurids. These running lizards, with sprinters' legs and serrated teeth, dashed about on two feet, each of which ended in a sickle-clawed, organ-stabbing toe, cocked like a switchblade. Biggest among them was *Deinonychus*, Greek for "terrible claw," which grew to the size of a large man and apparently hunted in packs. It hounded and harried the reptilian rhinos and giraffes of its day, slashing and ripping to death beasts much bigger than itself through gang strength.

It now appears that these terrible little pack lizards also came with feathers. Otherwise bringing nothing to bear on the predatory discussion, it's a bit of trivia worth contemplating while watching the mockingbird hunting beetles in the front yard. It also offers a hint as to what really became of the dinosaurs.

The Age of Mammals

Around sixty-five million years ago, a short time after an asteroid six miles wide struck the Caribbean basin, the dinosaurs disappeared. The errant celestial rock smashed a hole 3 miles deep and 112 miles across. At first it scorched, and then it chilled. Some say North America, with its *T. rex* and *Triceratops* and muggy conifer forests, was immediately fried. The rest of the world suffered more in the aftermath, as the sky darkened with ash and toxic gases, by some accounts chilling the earth in a shroud of shade. The connection between the cataclysm and the disappearance seems forever to be argued, but whatever the causes, the resulting cliff of extinction stands undeniably tall and steep.

Through the bottleneck of the Cretaceous-ending extinction, and into the void beyond, slipped the ancestral birds and the shrewlike mammals that had been scuttling in the dinosaurian shadows. And over the first few million years as the new crew in charge came out of hiding, new monsters began arising to make meat of them. From the ancient rainforests of North America, more than fifty million years ago, came one of the first apex predators of the age of mammals. It stood nearly six feet tall on two enormous legs, ran as fast as a deer, and gobbled primitive beasts the size of dogs. *Diatryma* had a battle-ax beak, an appetite for live meat, and wore feathers. It was a bird.

Diatryma had a tribe of counterparts in South America of similar design. Taxonomists classify them as phorusrhacoids; to most others they are the "terror birds." Though some have questioned *Diatryma*'s reliance on meat, the terror birds' massive hooked beaks and useless little wings left little doubt of their profession. These were birds to be imagined springing from the tall grasses and overtaking miniature horses and ancestral deer in open pursuit, seizing and beating their prey senseless against the ground before gulping them whole. One of the birds, a ten-footer named *Titanis*, once terrorized as far north as Florida.

In time the terror birds gave way to better, four-legged designs. By sixty million years ago, a distinct line of carnivorous mammals had appeared, the first of them in the shape of a weasel crossed with a cat. They were lithe, stealthy little predators, snaking through the undergrowth and tiptoeing through the canopy, each of them bearing a hallmark adaptation found about halfway back on the jaws. Opposing each other top and bottom, two large cheek teeth—later labeled the carnassials—bore cusps that had been honed to blades, coming together in a scissoring, slicing, meat-cleaving action.

Fore and aft, the carnivore mouth supplied a complete toolbox of the craft, leading the way with incisors for nipping flesh, followed by spiked canines for piercing and stabbing vital arteries and organs, ending in molars for gripping limbs and crushing bone. And invariably along the way there were those shearing carnassials. The teeth were set deeply in thick mandibles, the jaws levered by heavy temporal muscles attached to exaggerated ridges of skull bone. It was the carnivore's Swiss-army-knife alternative to the terror birds' basic maul of a beak.

From some such proto-carnivores arose nine major lines of meat-eaters, all but one still hunting today. They spread across the ecological spectrum, filling the land's top predatory niches. These were the ambushing cats and bone-crushing hyenas, lumbering bears and long-distance dogs. One line, on the way to becoming bears, split off and took to the water, feet morphing into the flippers of seals. Another line combined the strength of bears with the running mode of dogs to become the bear-dogs, a hybrid experiment lunging after hoofed prey across the ancient steppes of North America and Eurasia. From little slinking cats of Asia came the lion and tiger, rushing from cover and killing with suffocating throat holds. From North America grew a family of dogs, culminating size wise in the long-legged, distance-running, gang-tackling wolf.

In the seas, the missing mosasaurs were replaced by killer whales, hunting in packs capable of killing one-hundred-foot blue whales. In the skies over New Zealand flew an eagle with a ten-foot wingspan, chasing the island's famous flightless moas. On a few small islands in the Lesser Sunda of Indonesia, an ordinary monitor lizard grew to be the ten-foot Komodo dragon, credited with preying on prehistoric dwarf mammoths, and more lately on the occasional human. Australia did the Komodo one better, producing a land dragon named *Megalania* topping out at thirty feet and more than a ton.

Proven flesh- and bone-eating designs tended to be copied and reinvented. The hyena family experimented with bone-crushing beasts that could run like dogs. The dog family experimented with canid designs that could crush bone like hyenas. The daggerlike canines of the saber-toothed cat appeared in a variety of lineages across the ages. The wolf of the northern hemisphere had its counterpart in Australia's thylacine, a doglike marsupial more closely related to the kangaroo, which it hunted. The lion of Africa and Asia and prehistoric North America also had an Australian version weighing more than 250 pounds and raising its young in a pouch, like a possum.

The primates, for their part, sent forth from the savannas of Africa a superperpredator of their own, an odd and unimpressive line of bipedal apes bearing small teeth and no claws, and a curiously oversized brain. The versatile hominids married the group-hunting tactics of the wolf pack and lion pride with the scavenging skills of the hyena and jackal. The latest in the line came to be *Homo sapiens*, who within half a million years of his inception, had mastered the killing of the largest prey in the animal kingdom. And their predators too.

As recent as twenty thousand years ago, at least ten species of carnivores weighing one hundred pounds or more lived in North America. There were two species of wolf and three species of bear, one of which stood as tall as moose and ran with the speed of a quarter horse. There was *Smilodon*, the saber-toothed cat; an American lion, dwarfing its African siblings; the modern jaguar, puma, and an American cheetah. The continent was rich in superpredators. There were also, of course, supersize prey feeding them, mammoths and giant ground sloths most famous among them. Then quickly and mysteriously came their end.

By about thirteen thousand years ago, North America's suite of big predators had been halved. All the mammoths and sloths and three quarters of the largest hooved animals disappeared too. That they all went so very soon on the warming heels of a waning ice age—but so too on the arrival of spear-wielding hunters from Siberia—has since sparked one of the most enduring who-done-it debates of the last century.

By whatever cause, the great Pleistocene extinction had brought any escalating of predator and prey to an abrupt halt. But unlike others before it, this extinction was not followed by a revolution of wondrous new megabeasts. What followed was a pause of a few thousand years in which the continent's skeleton crew of survivors regrouped and settled into their new and hollowed-out surroundings. Yet that too was to be a brief interlude.

Where Wolves Once Ran

One October day of 1926, in a high and wild valley of grass and sage walled by the western spires of Wyoming's Absaroka Range, two young wolves followed their noses to the ripening carcass of a bison. In their hunger and haste, each of them stepped into a steel-jawed trap, which held them fast until their trapper arrived. By then about six months old, the pups presented two scared and sorry portraits of youth, feet caught in the cookie jar. The report doesn't specify whether the pups were dispatched with a bullet or the blow of a shovel head. They are recorded more simply by an old photo taken in the last minutes of their life, the two of them sitting worriedly upon the dead bison. They would be entered as statistics, the 135th and 136th wolves killed over the previous twelve years in that particular jurisdiction. Or, tallied in the grosser context, they were two of millions of fellow North American predators vanquished over the past century in an ardent cultural campaign to rid the country of them. For decades afterward, these two would hold the distinction as the last wolves born in Yellowstone National Park.

Over the years, reports would occasionally surface from somebody having just glimpsed another wolf in Yellowstone. None ever amounted to more than a transient loner hurriedly passing through. Yellowstone was no longer a place for a wolf to live. In that respect it rendered the first national park in the United States as hardly distinguishable from any other geographical address of the American West.

By the time poisoning was legally outlawed (but only partly curtailed) in the 1970s, gray wolves occupied less than 4 percent of their former range in the Lower 48. Its southeast counterpart, the red wolf, was already extinct in the wild.

By the time the wolves were driven from Yellowstone, the few still roaming outside had become a lonesome scattering of stragglers and freaks. They were the survivors of a countrywide population that a century earlier might have numbered a million or more. They were the endpoints of a campaign that had begun in 1630, when colonists of the Massachusetts Bay Colony were offered a penny per dead wolf. The U.S. government eventually made it federal policy to eradicate wolves and fellow vermin, with a tax-funded program in 1915 dedicated to the mission. With leghold traps and rifles to start, the United States added poison to its arsenal—thallium, strychnine, cyanide—and when aircraft became available, hundreds of thousands of little balls of fat laced with poison started raining from the skies. By the 1930s, most of the Lower 48 had recorded their last wolf. Those few remaining wolves were typically running around minus toes and feet lost to the jaws of steel traps, or carrying healed bullet wounds, mended bones, and memories of slain packmates.

The last wolves were plucky, cynical creatures educated in the frontier school of hard knocks, tiptoeing through a country brimming with bounty hunters and mined with traps. They were renegades who led the posses through months and years of cross-country chases and last-second getaways, tripping traps and stealing cattle on the run. These were seen as the lupine equivalents of battle-scared gangsters and gunslingers. They carried names like Old Lefty and Old Stubby. There was Three Toes of Harding County, South Dakota, and Three Toes of the Ashipapa, Colorado. New Mexico harbored a famous outlaw named Old Three Toes; Arizona had Old One Toe. Some of the wolves evoked supernatural powers of escape and resurrection. There was the Werewolf of Nut Lake, Saskatchewan, and the Ghost Wolf of the Little Rockies, Montana.

Those unfortunate enough to be taken alive were treated to a stupendous catalog of creative tortures. Captured wolves were scalped, lit afire, hamstrung with hunting knifes, bludgeoned with clubs, dragged to death behind horses and disemboweled by packs of hunting hounds. Survivors were those savvy enough thereafter to keep a foot beyond rifle range, paranoid enough to pass up any meat suggestive of strychnine, stepping clear of anything remotely smelling of human.

One of the most notorious of renegades was Lobo, who with his mate, Blanca, and phantom pack had run rings around the stockmen and trappers of the Currumpaw cattle range of northern New Mexico. Lobo and his pack were imbued by legend with monstrous size and speed, and a fantastic claim of having killed 250 sheep in a single night. By the time the flamboyant nature writer and erstwhile wolf-killer Ernest Thompson Seton was called in to take his shot, Lobo, the King of Currumpaw, had a thousand-dollar bounty on his head. "There was not a stockman on the Currumpaw who would not readily have given the value of many steers for the scalp of any one of Lobo's band," wrote Seton, "but they seemed to possess charmed lives, and defied all manner of devices to kill them."

Lobo's mate, Blanca, a white queen of a wolf, was the first to misstep into Seton's trap. Seton and his accomplice, not wanting to ruin the pelt with a bullet hole, used ropes instead. "We each threw a lasso over the neck of the doomed wolf, and strained our horses in opposite directions until the blood burst from her mouth, her eyes glazed, her limbs stiffened and then fell limp." Lobo fell shortly thereafter, held fast by a foot in each of four steel traps. The King of Currumpaw had been finally baited by the alluring scent of his lost mate, whose carcass Seton had shrewdly dragged atop the buried set of traps.

Inevitably even the legends grew old and gave out. By the 1940s, the last few representatives of Canis lupus in the Lower 48 had retreated to the far north woods of Lake Superior, or, in the spirit of Geronimo, had holed up in the mountain deserts of the Southwest—all clinging fiercely to the most impenetrable hideaways left within the country's borders. Wolves in respectable numbers still roamed the unpeopled reaches of taiga and tundra in northern Canada and Alaska. But as a breeding, pack-forming, hunting creature of any ecological consequence, the wolf had ceased to exist over the whole of the inhabited country.

The Last American Predators

That left two other animals from the top of the terrestrial food pyramid, neither of which fared appreciably better than the wolf. The grizzly, biggest terrestrial carnivore south of the polar bear, had with the waning of the Pleistocene ice come to occupy the western half of the North American continent, Alaska to Mexico, California coast to the Great Plains. And by the time those geographic coordinates were given such names, the great bear was in full-bore retreat.

Lewis and Clark, in their historic crossing of the country at the turn of the nineteenth century, met a discomforting thirty-seven grizzlies along the way, nearly every one of which they reflexively fired upon.

Some of the grizzlies returned the welcome. One bear took on six of Lewis's men and a hail of bullets. "Four . . . fired at the same time and put each his bullet through him, two of the balls passed through the bulk of both lobes of his lungs," Lewis wrote in his journal, "in an instant this monster ran at them with open mouth." Two more men fired, one breaking the bear's shoulder. The bear charged on. The men scattered, fleeing for the river. Two were able to duck into the willows and reload. "They struck him several times again but the guns only served to direct the bear to them, in this manner he pursued two of them separately so close that they were obliged to throw aside their guns and pouches and throw themselves into the river altho' the bank was nearly twenty feet perpendicular; so enraged was this anamal that he plunged into the river only a few feet behind the second man . . . , when one of those who still remained on shore shot him through the head and finally killed him." Upon hauling the great bear ashore, Lewis's men counted eight holes piercing its body in every direction.

"These bear being so hard to die rather intimidates us all," wrote Lewis. "I must confess that I do not like the gentlemen and had rather fight two Indians than one bear."

In the freshly blazed path of Lewis and Clark came the miners and trappers, buffalo gunners and ranchers, bringing cattle, sheep, repeating rifles, and similarly bad designs on the grizzly. By the mid-1990s, the U.S. bear range was reduced to half a dozen or so dots on the map, only two of them with animals enough to predictably locate them, namely the national parks Glacier and Yellowstone.

There had also been two big cats roaming the United States to be. The jaguar, El Tigre, the great spotted cat now largely confined to the tropical Americas, had ranged as far north as central California, and maybe as far east as the Carolinas, at the time the Spanish conquistadors marched ashore in the 1500s. The mountain lion—otherwise cougar, puma, or panther, according to regional dialects—could once be found from Atlantic to Pacific, sea level to alpine peaks, rainforests to deserts. By the midpoint of the twentieth century, its North American range had been halved. Supposed animals east of the Rockies became the periodic subjects of breathless sightings and perilously close encounters, but they somehow always escaped, leaving no pictures or other physical evidence of their existence. There did remain, however, a bona fide, if tenuous, holdout of eastern panthers, an inbred enclave numbering fewer than a hundred, precariously clinging to the ever-shrinking cypress swamps and pine forests of southern Florida.

The air too grew increasingly thin of predators. The golden eagle, with a seven-foot wingspan and an overblown reputation for killing livestock, made a natural target wherever it soared within range of a sheepman or fed-eral control agent. The aircraft added an unspoken element of sport to the business of shooting eagles. In the 1940s and 1950s, federal agents and sheepmen were gunning golden eagles by the thousands each year over the plains of western Texas and eastern New Mexico.

In the East, smaller raptors took their share of the heat. Every autumn thousands of hawks and falcons and a few straggler eagles flying south for the winter would funnel into one particular aerial highway of thermal up-drafts rising over the Appalachians of eastern Pennsylvania, to be blasted like flying rats. On one famous rocky outcrop now known as Hawk Moun-tain, squads of gunners would wait as the hawks streamed low overhead. Bodies would pile up. Gun barrels would overheat. On a good day's outing, the shooters would carpet their lookout ankle-deep with hawk bodies— some of the wounded still flopping—and proudly take pictures.

Beginning of the End

The American history of predator eradication is notable mainly for its breathtaking rapidity, though not for its novelty. In terms of general pattern and process, it's a story as old as civilization, arising some ten thousand years ago in the valleys of the Tigris and Euphrates of Mesopotamia and the Nile of Egypt, with the advent of livestock.

There, in the Fertile Crescent, the first systematic attacks on the great predators was launched. They followed in logical step with the first hunters and gatherers who took up the experiment of farming in the cradle of civilization. They cultivated the local barley and wild wheats. They corralled and tamed their wild game. From the spirited mountain populations of wild mouflon sheep and bezoar goats came their docile, domestic barnyard counterparts; from the ornery aurochs—inspiration of the man-eating Minotaur—came the placid, cud-chewing cow.

For the prototype herdsmen, these were intended as handy packages of fresh meat. The wild carnivores with which they shared the land quite naturally tended to see them that way too. With domestication were born two new and related concepts: one viewing animals as property, the other assuming predators as thieves and threats to livelihood. The inevitable conflict was compounded as the proliferating humans and their herds expanded across the countryside, chopping forests, cultivating fields, grazing the fields, supplanting the wild herds as they went—in essence making livestock the only game in town for a big hungry predator.

And so it was that by the end of the Stone Age there was a price on the predator's head. Solon, a noted statesman of Athens twenty-five hundred years ago, announced a five-drachma bounty for anyone killing a wolf. The Celts, for their part, had by the third century B.C. bred the massive wolfhound, sending it bounding in pursuit of wolves over the Irish countryside.

At least one other force of reason sufficed for eliminating the large predators. Killing them was such high entertainment. With the advent of crowded cities and stultifying labor, there arose a class of citizen inevitably bored with the sedentary life, whose legions could apparently be appeased by making live theater of animal slaughter. With the raising of their great amphitheaters, the Romans began hauling elephants, hippos, and lions by the thousands, from as far away as Mesopotamia and Africa, to be killed in grand spectacle. A single day's slaughter in the Coliseum of Rome might include a hundred bears, more than four hundred leopards, and five hundred lions—quantities better understood when considering that a take of five hundred lions would more than wipe out the entire population now living in Asia.

By medieval times, predator persecution had gone large-scale, most conspicuously in the pursuit of the wolf. There were snares and pitfall traps and stomach piercers, as well as the classic steel foot traps. Firearms debuted, armies amassed. Whole villages would turn out to drive wolves into nets or shoot them as they broke back through the lines.

With the persecution rationale and weaponry in place, the big predators toppled across the civilized reaches of Eurasia. From Scandinavia to the southern Alps, wolves and bears and lynx were driven to the craggiest mountain enclaves, or eastward to the most isolated reaches of Siberia.

In Africa came more numbing repetition. The African wild dog—Africa's pack-hunting version of the north's gray wolf—was once a cosmopolitan predator of sub-Saharan Africa; by the end of the twentieth century it had been chased to a scattershot of guarded reserves.

The cheetah, by the mid-twentieth century, was all but driven from the Asian continent, all other populations collapsing toward the last big parks of Africa. The lion, archetypal African predator, descended down a similar track: Its widespread presence in western Asia was whittled to some 350 individuals confined to a single forest in India; its supposed stronghold in Africa was a façade, with populations heading for impending collapse beyond the confines of guarded parks and reserves. The tiger, with its pan-Asiatic range, was evaporating to a handful of disparate populations, the whole of them numbering maybe seven thousand.

As in America, the international pogrom visited upon the terrestrial carnivores extended logically to their aerial counterparts. By the 1960s, the patterns were in place that by the end of the century would relegate more than forty species of the world's raptors to the International Council for Bird Preservation's list of threatened species.

The slide of top carnivores' continued into the sea, where the big marine predators were something more than competitors for food; they were the food.

The Atlantic cod, New England's founding fish, the currency of northern Atlantic nations, once swam in great schools of six-foot monsters, in numbers so thick that John Cabot in 1497 bragged about landing them with a bucket lowered into the sea. After the buckets and hooks were replaced in the 1920s with trawling nets, the cod began to disappear. By the 1960s, cod stocks and cod economies were collapsing, from the seas of northern Europe to the Grand Banks of Canada to the Gulf of Maine. A big cod now, in the few places they are still numerous enough to be fished, averages a little more than a foot long.

The Atlantic bluefin tuna, a half-ton, hot-blooded torpedo hitting speeds of fifty miles per hour, is a creature that crosses oceans as a habit, chasing down smaller fish in a violent rush of muscled propulsion. It is also a fish that often ends up as frozen slabs of prospective sushi in the massive fish market of Tokyo. A single prime bluefin will sometimes fetch sixty thousand dollars at the market. Thirty years ago, there was a breeding stock of about a quarter million bluefin tuna in the western Atlantic. Today they number about twenty-two thousand animals.

And so it goes, on down the line of the ocean's biggest fish. Whereby the fall of the great terrestrial predators can be summed up as a casualty of the agricultural age, the subsequent collapse of their marine counterparts owes itself to the coming of the technological age. Following World War II, the machinery of war was redirected at the ocean's fishes. By the 1960s, the fishing business entailed tracking schools by radar and satellites, shooting harpoons with howitzers, and spreading nets wide enough to swallow jumbo jets. To be a big fish in the 1960s was to wear a bull's-eye.

In 2003, two Canadian scientists confirmed the gut-level suspicions of observers around the world. In a paper published in the journal *Nature*, Ransom A. Myers and Boris Worm exhaustively tallied the catch of the world's tuna and sharks, marlins, swordfish and codfish—the oceans' alpha fish—starting with the industrial fishing revolution of the 1960s. They conservatively estimated that the numbers of the ocean's top predators were down by 90 percent. "From giant blue marlin to mighty bluefin tuna, and from tropical groupers to Antarctic cod, industrial fishing has scoured the global ocean," said Myers. "There is no blue frontier left."

Upshot

So how are those big predators doing, anyway? By the time that question had gained scientific legitimacy—by the time that mischievous trio Hairston, Smith, and Slobodkin had tossed their firecracker into the marble halls of ecology, declaring the world green on account of predators; by the time their protégé Bob Paine had returned from the tidal rocks of Mukkaw Bay bearing important news of a certain sea star named *Pisaster*—all but the wildest outposts were in the terminal stages of apex predator annihilation.

In place of Paine's pitching arm were poisons and guns and automobiles, evaporating wildlands and shrinking prey populations, removing the last of the top predators wherever they lived, gray wolves to white-tipped sharks, mountains to plains, coasts to ocean depths.

It was a planetary experiment of predator removal, perversely reminiscent of Paine's own trials upon the coastal rocks, but with two critical departures. To gauge the impact of *Pisaster*'s absence, Paine had dutifully set aside a length of shore untouched and still crawling with starfish—a benchmark of relative normalcy—a control. On that account, the global experiment fell decidedly short. For the world's alpha predators, there was no inviolate refuge of rock. There was just one ubiquitous spread of civilization, chucking as they went every last starfish into the sea.

The question of how much the big predators mattered had become a problem almost impossible to measure scientifically. If one were to truly begin testing for the consequences of their disappearance, it would take a study area of formidable size. It would require funding on a scale better suited to the space program than the threadbare coffers of the National Science Foundation. It would require either moving or killing a lot of big and beautiful and rare animals—if such might still be found.

With the environmental awakening and endangered species legislation of the 1970s, it was the sort of experiment that would have been rapidly shuttered out of ethical principle, if not by threat of imprisonment. And so it was a fortuitous miracle that by the 1970s, the predator-prey experiment an ecologist could only dream of was already three hundred years under-way, waiting to be discovered.

THREE   Forest of the Sea Otter

Discovery

In 1741, the Siberian sailing ship St. Peter, commanded by Vitus Bering and his crew of seventy-seven, got lost while exploring the uncharted western shores of North America. Battered in the cold, tempestuous sea, its sailors wracked with scurvy and desperate with hunger and thirst, the St. Peter made way back to the Kamchatka Peninsula until running aground on a bleak volcanic island at the western end of the Aleutian archipelago. As castaways go, the crew of the St. Peter could have done worse. The shores of their shipwreck were swimming with easy meat.

All about the island appeared fur seals, which, when not arcing through the surf, were napping on the beaches. There was a peculiar cormorant, a fearless and flightless swimming bird the approximate size and flavor of a roasted goose. Grazing its way through the seaweed shallows was an unimaginable grublike mammal growing thirty feet long, propelled by a tail fluke. Steller's sea cow, named after the ship's naturalist, was slow to flee, easily gaffed, and insulated with a nine-inch layer of buttery fat that Steller himself likened to "the oil of sweet almonds."

Amid these beds of floating kelp also swam sea otters, four-foot-long slips of shimmering fleece with button eyes peeking from furry, quizzical faces. When not diving for shellfish, the otters tended to recline upon the water like logs of driftwood. Sometimes they would haul out and sleep atop the rocks, allowing Bering's crew to tiptoe down and club them.

Bering's wreck would likely have earned but a footnote in history had it not been for the sea otter. Their meat was reported to be "tough as sole leather," but their fur was softer than the finest silk. Sea otter fur, constructed of hairs clustered 645,000 to the square inch, served the otter in trapping an insulating layer of air against the body, one of the thermal innovations allowing this blubber-free animal to make a living in the blood-chilling waters of the Bering Sea. (The other being the metabolism of a blowtorch, fueled by an appetite consuming a quarter of the otter's body weight each day.)

Of course none of these physiological trivia mattered to an eighteenth-century explorer. What mattered was that this was the thickest, plushest, most decadent design of any mammalian coat. As it happened, a certain strata of Chinese aristocracy had by that time developed a number of other designs for sea otter fur, favoring it in the form of capes and sashes, mittens and caps, as well as the trim of fine silk gowns. It was a market to make a poor enterprising Russian rich. When the St. Peter finally limped home to Siberia, jury-rigged and lightened by the death of Bering and more than half his crew, word of the otters went with it. Little more than a year later, a stampede of maritime gold diggers was sailing a beeline from Siberia for the mother lode of Bering Island.

To be a naïve fur-bearing mammal or a fresh piece of meat lounging in the path of the promyshlennik—as the Russian fur hunters were called—was most unfortunate. Within twenty-six years, Steller's sea cow was gone, never to be seen again. The cormorant, now known in history books as the spectacled cormorant, held out until about a century later.

The fur seals, by strength of their mobility and sheer multitudes, maintained a slim and sporadic presence through the slaughter. But the shore-hugging sea otters—so conveniently gathered, so rabidly assailed—evaporated wholesale through the Aleutian chain, one island after the next. With club and gun, net and spear, and an army of Aleut slaves drafted to the task, the promyshlennik swept across the Pacific, killing otters in the water, asleep on the rocks, pups and nursing mothers all. America and Europe inevitably came scrambling for their piece of the action. And together the nations raced through the Gulf of Alaska and down the coast of North

America, through British Columbia, Washington, and Oregon, California to Baja, chasing otters to the last.

A treaty in 1911 brought a figurative cease-fire, given by that time nobody could find otters enough to hunt. In the century and a half since the killing had started, anywhere from half a million to nine hundred thousand sea otters had been mined from the Pacific.

Yet somewhere, somehow, slipping cautiously through a few overlooked coves and sheltered reaches of rocky coast, a dozen or so tiny pockets of sea otters survived. From their secret sanctuaries they began to rebuild. As the sanctuaries filled to capacity, hungry pioneers began crossing ocean gaps, recolonizing emptied shores. By the late 1960s, there were again big pods of sea otters to be found bobbing in the cold Pacific swells. They had gone part way toward refilling the ocean-wide void left by the fur rush. And in doing so they had set the stage for another revolution of sorts, this one destined for the history books of ecology.

Estes

It was about that time that a graduate student fresh out of Washington State University was flunking his military physical. James A. Estes was a young man with a new master's degree in zoology and a sense of resignation that he would soon be stowing it indefinitely for a tour of combat in Vietnam. At the qualifying physical, the examining doctor sailed through the procedure, then in closing probed Estes for any unseen problems. Once a promising pitcher in college, Estes casually mentioned the shoulder that had given out and ruined his fastball.

Anything else? asked the physician.

Well, there's this knee that tends to pop.

Can you show me?

Estes shrugged and tweaked the knee, detonating a minor explosion.

"The doctor's eyes went big," said Estes, "and that was it. He rejected me right there."

And so the slim young man with the bum shoulder and trick knee suddenly found himself with the unanticipated freedom to do whatever he wanted with the rest of his life. At which time the phone rang, with news from a former professor about some island called Amchitka.

The island had been a strategic outpost in the North Pacific theater of World War II. And in the cold war that followed, there had sprung up a bustling little enclave of barracks and trailers, hard hats and physicists of the Atomic Energy Commission, preparing to test the island's reaction to a series of underground nuclear explosions, the biggest of which was to produce four hundred times the power of the Hiroshima blast. The testing included an environmental branch of operations. And from that branch, there was some money and an opportunity for researching something or other about sea otters, with which Amchitka by this time once again abounded. Any interest?

Estes was soon thereafter packing for the North Pacific. He had established his academic base camp as a Ph.D. candidate at the University of Arizona. (Arizona might have seemed a long way from the Aleutians, thought Estes, but then again, where on Earth wasn't?) And in 1970, with his marine biology résumé consisting of little more than a fondness for fishing and a tendency for seasickness, he shipped off to one of the most alien shores this side of Vietnam to study sea otters.

Up to that point, almost everything Estes knew about the Aleutian sea otter had come from a single definitive monograph published only the year before by a federal wildlife biologist named Karl W. Kenyon. Kenyon had spent a decade watching otters from the bays and rocky promontories of Amchitka Island. He'd noted the daily schedules of these creatures of habit, of busy mornings and late afternoons spent diving for food, gathering mussels and urchins and fish from the rocky bottoms, interspersed with afternoon naps and bouts of fastidious grooming.

The sea otter is the largest and most seaworthy member of the carnivore family of weasels and badgers called the Mustelidae. Nearly as supple as a seal in the water (and with its enormous webbed feet, nearly as clumsy on land), the otter dives with an undulating body propelling it mermaid fashion to bottom depths as great as eighty feet. In dives averaging a minute to ninety seconds, the hunting otter gathers as it goes, grabbing prey with its forepaws, stuffing its baggy armpits full of food, then popping to the surface to set the table. On its chest the otter places its urchin—or mussel or fish or any of the 150 species of prey on its menu—and begins eating. If a shell is too tough to bite through, the otter pulls a rock from its stash, props it upon its chest as an anvil, and with its paws hammers the shell into submission. Sea otters tend to eat whatever comes most readily, an otherwise intuitive habit that would come to bear greater significance in matters of ecology.

Touching down on the runway of Amchitka, Estes wondered just what he had gotten himself into. Amchitka arose as a spike of rock in the far western reaches of the Aleutian archipelago, near the end of a twelve-hundred-mile volcanic arc thrusting like an elephant's tusk from the Alaska Peninsula toward Siberia. For one raised in the sunny chaparral suburbs of San Diego, the gray, gale-whipped waste of the Bering Sea was the scariest place imaginable. And what about those odd, aquatic carnivores, those eighty-pound sea weasels Estes was supposed to make sense of? Maybe he'd research something about their natural history, or their ecology—perhaps connecting them somehow with their coastal ecosystem might be an interesting thing to do. Sea otters seemed like a cool thing to study, thought Estes, and that was about it.

Estes spent much of the next year acquainting himself with his subjects, netting and tagging and following the sea otters of Amchitka, to what end he still wasn't sure. He started scuba diving to get a better look at where the otters spent a good part of their lives. What he saw confirmed much of what Kenyon had surmised before him: that otters floated and dove, slept and mated and died among the marine forests of kelp rooted in the cold coastal waters. Estes knew the otters had a catholic diet, but that many had a particular fondness for sea urchins. Otters, urchins, and kelp. He had no idea how they all might fit together. He wasn't yet sure what his question was. He wasn't aware that half an ocean away, in a crucible of ecological discovery called Mukkaw Bay, another scientist was busy formulating the question for him.

The Godfather

By the late 1960s, Bob Paine was off and running on the rocky shores of Washington. With his landmark study linking the predations of Pisaster to the intertidal diversity of species, Paine had realized in Mukkaw Bay a perfect little laboratory for asking some of the biggest questions of ecology. The splash zones and tide pools of the rocky outer coast offered a dynamic universe of predator and prey in a fishbowl, which Paine and his lengthening list of student protégés could poke and prod and manipulate as they pleased. There were neither rare nor endangered species to tiptoe around. In this Lilliputian kingdom there was no creature so fleet or fierce as to run away in the middle of an experiment, or rear up and maul the probing naturalist. Nobody cared if Paine drilled survey stakes that would benchmark a study plot for decades. If for experimental purposes he needed a new tide pool, he could jackhammer one. Nobody would object if, in the name of science, he pried a whole galaxy of starfish from a wave-beaten rock and flung them into some other orbit.

And if, say, Paine might be curious to see what would happen if he plucked all the sea urchins from a particular tide pool, what was to stop him? So Paine, master of his universe, did so. Along with student Robert Vadas, Paine exercised his godly powers on an ecosystem dominated by Strongylocentrotus purpureus, the purple sea urchin. The urchins are globular, spiny cousins to the starfish, centered underside with a circular jaw used for grazing kelps and other algae off the bottom. They are to ribbons of kelp as the starfish are to mussels. When Paine had first peered into the tide pools of Mukkaw Bay, many had been cobbled solid with sea urchins, the kelp hardly to be seen. Within a year of he and Vadas playing predator to the urchins, the once neatly shorn pools had converted to solid jungles of fleshy kelp.

The resulting paper, a minor classic in Paine's growing canon, presented yet another stark example of an animal's consuming potential to upend an ecosystem. This was Paine's forte, a knack for drawing grand panoramas from intimate little portraits of life. The discovery that a sea urchin could obliterate a tide pool of kelp was maybe not as newsworthy as that of an earth-bound asteroid. But then again, kelp was something more than just seaweed.

Kelp is a generic term for a suite of brown algae species that dominate the rocky shallows of the world's cold-water coasts. When washed up on the beach, they are those familiar squishy, translucent brown straps of lasagna pasta, pecked about by the crabs and gulls. When attached in living colonies to the seafloor, kelp is a different animal, a lush and leafy marine forest. The giant kelp Macrocystis pyrifera, the sequoia of the kelp forest, can grow nearly two feet in a day, topping out at almost two hundred feet. Within the kelp forests congregate masses of marine life, from sea squirts to sea lions. Its photosynthetic flesh provides pasture for the swimming herbivores, its dead fronds fall as detritus for the bottom-scrapers. Its physical presence—its roots and trunks, its spreading canopy of fronds—is a scaffolding for life in an otherwise gripless realm of open water. For its unparalleled productivity and physical stature, the community of kelp has been seen by more than one ecologist as the marine analogue to the terrestrial rainforest.

"The number of living creatures of all orders, whose existence intimately depends on the kelp, is wonderful . . . I can only compare these great aquatic forests of the southern hemisphere with the terrestrial ones in the intertropical regions," wrote the young Charles Darwin on June 1, 1834, while passing through Tierra del Fuego on his five-year voyage aboard the H.M.S. Beagle. "Yet if the latter should be destroyed in any country, I do not believe nearly so many species of animals would perish as, under similar circumstances, would happen with the kelp. Amidst the leaves of this plant numerous species of fish live, which nowhere else would find food or shelter; with their destruction the many cormorants, divers, and other fishing birds, the otters, seals, and porpoises, would soon perish also; and lastly, the Fuegian savage . . . would . . . perhaps cease to exist."

Kelp was important, and urchins ate kelp. That much Paine had come to understand when, in the summer of 1971, faculty heads at the University of Washington sent their emergent star of intertidal ecology on an Aleutian junket. Paine was to pay an advisory visit to some UW graduate students then poking about the kelp forests far off on the island of Amchitka. One of those students was John Palmisano, who alerted his colleague from the University of Arizona that there was a terrific ecologist on the way. You really ought to meet him, Palmisano told Estes.

A Summit of Two

There's a small disagreement between Estes and Paine about where they actually met. Estes says it was waiting for a movie to start at the Amchitka gymnasium. Paine thinks it was in a bar. Whichever the venue, the ensuing chat inarguably altered Estes's life.

Estes was intimidated by this towering figure of science, articulating in bold, baritone clips of plainspoken English his views on how the world of nature worked. But Paine turned out to be an enthusiastic listener as well, sitting attentively as Estes offered his study plans. He listened until Estes finished explaining that he thought it might be instructive to show how the kelp forest supported its sea otters—ecology's conventional bottom-up line of inquiry.

"Jim, that's stupid," Paine said.

And in an act of big-brotherly affection, Paine flipped Estes on his head, conceptually speaking. "Jim, there's something vastly more interesting. You're sitting on a gold mine of interaction. You want to work on otters eating sea urchins, and sea urchins eating kelp," declared Paine, punctuating his advice with references to a certain paper recently authored by Paine and Vadas, verifying the urchin-kelp connection. Paine had seen what urchins could do to a community of kelp when left unchecked. And sea otters ate sea urchins, that much Estes certainly knew. Here, then, said Paine, was a ready-made experiment of predation on an oceanic scale, a checkerboard chain of islands, some with sea otters and some without (compliments of the fur hunters), waiting for somebody to take its measure.

Estes didn't remember much of the movie, dizzy as he was from having his worldview turned upside down, compliments of Robert T. Paine.

Shemya

The idea was to go to another place with all the ecological qualities of Amchitka, but lacking otters, and compare the two: Go underwater, count urchins, measure the kelp, and note any differences the sea otter made. If one could just see any significant changes in urchins, thought Estes, that would be exciting enough.

Shemya was the first choice. Closer to Siberia than America, Shemya was six square miles of rock protruding more than two hundred miles to the west of Amchitka. Shemya, like Amchitka, had a lingering military presence, with an air force base and a runway. It had but a small pocket of otters that had managed to recolonize since the great extermination. The rest of the island would provide the counterpoint to the otter-laden Amchitka.

In 1971, Estes and three colleagues arranged a reconnaissance, chartered a flight, flew the gap from Amchitka, and spent their first half day in Shemya under house arrest. There they stayed until the commanding officer finally softened to the idea that the four young hippie types appearing before him from out of the Pacific blue were not spies. Upon release, the reinstated biologists wandered down to preview the beach where they were to begin their surveys the next day.

The first things Estes noticed were the sea urchins. Their skeletons littered the shore. He had never seen such big urchins at Amchitka. But here lay monsters, huge urchins four to five inches across, piled in green windrows at the high-water mark. This was something radically different. That night, Estes declared to his research mates, "We're going to see something cool here."

The next day they took a boat out onto the reef. Estes donned his wet suit and dove underwater, into a moment forever crystallized in his mind. "Because there it was," he said, "just this green carpet of sea urchins everywhere. And no kelp. And boom, that did it. It took me about a nanosecond to connect on what was going on with otters and the kelp forest system, and how important they were."

Here was a seascape that had been mowed to the nub—the antithesis of the Amchitka kelp jungles. And the only apparent difference was the sea otter. The most fertile ecosystem across the North Pacific coast was denuded for lack of otters. If what Estes had seen held true across the board, then the foundation of this coastal food chain was ultimately governed from above, by an adorable, urchin-munching carnivore.

Back at the barracks that night, Estes sat awake with his journal, writing and writing, frantically until dawn. Scraps of thought and observation bubbled up, gathering like magnetic filings into a pattern of life in the kelp forest. Otters to urchins to kelp, a chain of ecological percussion and repercussion, transforming an undersea world. Estes hurried to capture these ideas. He didn't dare chance that they'd abandon him in his sleep.

That one moment at Shemya had lifted the last of the fog. Estes now knew exactly his path. He set about measuring the sea otter's impact with scientific precision.

Each day, he and Palmisano would either wade from shore or head out in the skiff and then dive to the bottom of the reef, counting sea urchins and estimating the coverage of kelp. To Shemya's study they later added the otterless island of Attu, thirty-five miles farther out on the tip of the Aleutian chain.

Their data did not come cheaply. The two typically headed out into the Aleutians' notoriously deadly waters whipped by tireless gales blowing fifty knots. On days when even the locals hunkered on shore, Estes and Palmisano were out on the high seas. And Estes, the diver, was in the water. He dove both summer and winter. In the winter the waters averaged thirty-five degrees Fahrenheit; in the summer they warmed to forty-five. Estes would dive in the morning for an hour, collecting samples of kelp and urchins, staring into square-meter quadrates, filling data sheets with names and numbers. He would write underwater on slates until his chilled hands were shaking too violently to continue. After a hot shower to thaw the blood, he would head out again in the afternoon for another hour, until he was too numb to write anything more. The next day he would repeat the process.

It was the scientific way of objectively confirming what the eye could not deny. Estes and Palmisano had watched the defining event thousands of times at Amchitka: sea otters, lying in the water, bellies to the sky, urchins between the paws, rolling them like little balls of dough to flatten the spines, and popping them down the hatch like popcorn, one after the other. The inquiring scientist didn't need to sift through buckets of sea otter scat, didn't need to slice into stomachs to know what the otters were eating. And it only took a minute underwater to confirm the ecological consequences.

Wherever they looked, the pattern held true. On the otter-patrolled reefs of Amchitka there were forests of brown kelp, with fish swimming among the rising fronds, and a colorful diversity of sponges and hydrocorals, mussels and barnacles populating the seafloor. In the otterless reefs of Shemya and Attu the seafloor had been reduced to a pavement of pink coralline algae, pocked with spiny green blobs of enormous sea urchins. If Amchitka was a kelp jungle, Shemya and Attu were urchin barrens. It was every bit as obvious and dramatic as stepping from a towering ancient forest into clear-cut. And the sea otter remained the difference.

As obvious as it might have seemed—after all, it was no secret that otters ate urchins, that urchins ate kelp—still nobody before Estes had connected the first to the last. Karl Kenyon, in his decade with the Amchitka otters, had come close. "Indirect evidence from Amchitka Island, where a large sea otter population exists, indicates that sea otter predation has drastically reduced certain food species there," Kenyon wrote in his 1969 monograph. "Small green sea urchins are abundant. It is not possible, however, to find large individuals in the intertidal zone and I seldom saw an otter eating an urchin that approached in size the large individuals which are abundant in other Aleutian areas where the sea otter is scarce or absent." Kenyon, who did not dive, never mentioned the kelp.

By the end of their second summer of diving, Estes and Palmisano had the data and the confidence to go public with their findings. The two spent a week writing up the manuscript. Estes wasn't sure anybody would be interested, but Palmisano suspected otherwise and insisted they send their manuscript to the journal *Science*. And in September 1974, the most prestigious of scientific publications pictured Estes and Palmisano's sea otter on its cover.

Estes, who a few weeks before had been just another grad student wondering how he was going to survive in the world, was suddenly entertaining University of Arizona faculty who'd never before said a word to him, now asking for reprints. Letters came from afar. Professors had a different look on their faces when they passed Estes in the halls. Even the modest mind knew that this discovery was a big deal.

Estes and Palmisano had inadvertently added another brick to the infamous green world hypothesis of Hairston, Smith, and Slobodkin, whose paper had sensitized—if not irritated—so many ecologists to the notion of predators having so much to do with the state of the planet. The sea otter's kelp forests were in fact brown, but the analogy was apt. At least from a coastal Aleutian perspective, the world was brown because sea urchins don't eat all the kelp, because sea otters eat sea urchins. The sea otter had triggered a chain reaction across two links in the food web, the most compelling example since *Pisaster* of what Paine himself would later famously coin as a "trophic cascade."

Do predators matter in the web of life? They certainly did on the rocky shores of outer Washington state; and now, across a continent's breadth of North Pacific kelp forests, there was a compelling case for an urchin-eating carnivore whose presence could flip an ecosystem like a switch.

The *Science* paper was soon succeeded by a string of correspondences from the Aleutians, lauding the kelp forests' furry new patron saint of biological diversity. In 1978, Estes, along with his doctoral adviser, Norm Smith, and colleague Palmisano, emphatically reaffirmed the sea otter's top-heavy ecological influence with a paper in the journal *Ecology*. In it they painted an even more compelling picture of life with otters on the Aleutian reefs. Where there were otters, there were forests of kelp, and where there was kelp, there were extraordinary gatherings of fish. There were fish feeding on kelp, hiding in kelp, chasing smaller fish in kelp. On the tails of the fish came harbor seals, in numbers conspicuously elevated beyond those in the urchin barrens of Shemya and Attu. Bald eagles, feeding largely on fish and seabirds as well the occasional baby seal and sea otter pup, were abundant at Amchitka and nearby islands, absent on Shemya and Attu. The life energy flowing from the gardens of kelp was forever tracing back to this one adorable little carnivore.

Estes's otter had become the new textbook example of a keystone predator, eclipsing Paine's less cuddly *Pisaster*. It had emerged—as Paine himself would come to praise it—as "the poster child of marine near-shore ecology." It rang all the right bells—charming, charismatic animal hero with the tale of the comeback kid, bearing great ecological powers and a vital conservation lesson: Save the sea otter, save the forest. It was the stuff of instant legend. And it would have made for a much happier story had it ended right there.

FOUR The Whale Killer

Novelty ordinarily emerges only for the man who, knowing with precision what he should expect, is able to recognize that something has gone wrong.

—**Thomas S. Kuhn**

Hatfield's Observation

On June 10, 1993, an assistant of Jim Estes's was training his telescope upon the coastal waters off the island of Amchitka. The sky was standard Aleutian gray, the bay water blessedly calm for the purpose of observing wildlife. Gulls cried and eagles chittered overhead. Into the viewfinder appeared the black dorsal fins of a pod of killer whales. And completing the Bering Sea diorama, a furry-faced sea otter lay floating amid the canopy of kelp.

These were good times to be a sea otter. Or one who studied them. In the twenty years since Estes and his crews had started taking measure of the Aleutian coastal ecosystem, the otter population had prospered, filling in the fur hunters' century-old gaps to the point of overflowing. The Amchitka otters in particular had reached that mystical pivot point in ecological theory known as carrying capacity—the point of saturated abundance beyond which starvation and sickness would be expected to begin weeding the surplus. Here, thought the ecologist Estes, by then a professor at the University of California at Santa Cruz, was a rare opportunity to observe a top predator at the height of its profession, unbound from persecution, operating at the limits of its food supply.

On this afternoon, it was Brian Hatfield's turn at the watch, glassing the goings-on of the Amchitka kelp beds. The otters were mixing work and leisure as usual, gliding through the smooth water, fastidiously grooming the plushest fur in the animal kingdom, reclining with lunch on the chest. And there in the background swam those killer whales, four black-and-white behemoths filling the scope, one with the towering dorsal fin of a grown male. It was not unusual to see killer whales in these waters. Moreover, in recent years they'd become especially commonplace to Estes and his crews, who would see them two to three times a day in their otter surveys, often up close and lolling along the edge of the kelp beds. It was a curious new wrinkle of killer whale behavior to which nobody had given much thought.

At 1400 hours, as noted in his journal, Hatfield startled to a resounding splash about two hundred yards out. He scanned to find the big male, and at the leading edge of its wake, racing for the kelp bed, a sea otter. There was something strange about the otter, even as it reached the safety of the kelp. Back and forth the otter swam in spastic disarray, its hind flippers flailing the air. Hatfield, in his thousands of hours behind the spotting scope, had never seen such behavior. He had never seen what he could only surmise was the tortured writhings of a sea otter wounded by a killer whale.

Hatfield continued watching the queerly swimming otter, but now with another eye tracking the killer whales as they cased another bed of kelp. He noted, at 1430 hours, about three hundred yards from the scene of the attack, the dorsal fin of a whale slowly cutting toward a sea otter. The otter lay until the whale was twenty feet away and then dove. As it surfaced, the killer whale was waiting.

Into the kelp the killer whale charged. The otter doubled back and dove, resurfacing behind the whale. The whale turned about; the otter dodged again. Hatfield recognized the otter's tactic as one commonly used on him while he tried to capture them from his skiff. The otter made a final lifesaving dash for refuge in the rocky shallows, where the whale gave up the chase.

That night at the Amchitka mess hall, Hatfield relayed the events to his boss. Estes knew Hatfield as a keen and dependable observer, but this had

Estes thinking twice. "I don't know, Brian," said Estes. "I've been watching these things for years. And I've never seen a killer whale attack them."

Little more than a year later, another of Estes's team came to him with a similar story. On yet another stretch of Amchitka shoreline, Tim Tinker had noticed three dorsal fins circling an otter before one of the whales lunged out of the water and came smashing down. Whale and otter disappeared; only the whale resurfaced. Still blinking over what he had just seen, Tinker followed the whales as they approached a second otter. Again the dorsal fins circled, and again a killer whale came crashing upon an otter that Tinker never saw again.

Estes didn't know what to imagine. After two decades among the Aleutian otters, he had come to believe he could trace their food chain as easily as his way to the Amchitka mess hall. Suddenly there's a six-ton killer whale lying broadside across his path.

It was a phenomenon weird enough to wonder about, but by that time there was a more worrisome anomaly that begged more immediate attention. There were hints that the Aleutian otters, now half a century into their celebrated recovery from the brink, might in fact be crashing once again.

Collapse

In 1994, work on the Amchitka otters was coming to a close. The grant money had expired, the navy had pulled out (taking along its logistical support), and the otter research team was preparing to trade the ghost island of Amchitka for another windy gray outpost to the east called Adak. Before leaving, they conducted one last census of Amchitka's otters, scanning the island's entire 120-mile perimeter, as Estes had done more than twenty years ago in his inaugural explorations. Estes had counted upward of seven thousand otters in the early seventies. Now the count came back three thousand short. Upon looking around, the decline appeared to have legs. A more extensive, Aleutian-wide aerial census conducted just two years earlier by the U.S. Fish and Wildlife Service had tallied a 50 percent drop in otters since 1965. The numbers at first surprised Estes, though not stunningly so.

It was still possible, he thought, that the shortfalls lay within the range of experimental error, or the realm of natural fluctuations.

It was hard to imagine otherwise. By then the otter's legend had been securely enshrined in the textbooks of ecology, thanks to Jim Estes. With the fur hunters long departed, it seemed the only question left was how many of these benevolent predators the system could hold. The idea that the Aleutians should be anything but a full house of sea otters just didn't want to enter the mind.

The trouble followed Estes to Adak, where in Kuluk Bay two of every three sea otters disappeared in their first year under watch. Now the paltry census from Amchitka, along with the falling aerial counts across the Aleutians that Estes had initially soft-pedaled, began to strike a more serious chord. Estes had been watching these otters long enough to understand that something was fundamentally wrong. But when it came to recognizing what that something was, perhaps he'd been watching too long.

His assistant Tim Tinker was not yet so encumbered by the prejudice of experience. Tinker had just begun observing the otters of Adak in intensive spells, recording their feeding habits, tracking their movements. Every month or so, he would also note a pod of killer whales come by. And a day or so later a couple more of his tagged otters would be missing. The coincidences began to click. Tinker had also been the one watching in Amchitka a year earlier, when the two otters were body slammed and presumably swallowed in order. When in November 1995, four more of the otter team stood and watched as a killer whale dispatched an otter—this time at Adak—Tinker could no longer deny the conclusion that for Estes was still unthinkable.

Tinker phoned his boss with his killer whale theory for the missing Aleutian otters.

My god, thought Estes. He's been up there too long.

But how else to reconcile the evidence? They had run the blood tests, measured the body fat—there were no signs of disease, contaminants, or starvation to explain the diving numbers. The otters of Adak were pumping out pups at healthy rates. Most confounding of all, there were no bodies. As a general rule, dead otters do not sink. Every winter during their years at Amchitka, Estes and the otter crews would walk the beaches, ten to fifteen miles at a stretch, crossing dozens of carcasses and skeletons washed ashore—the sort of natural attrition to be expected for a population too crowded for their own well-being. But after two years of walking the beaches of Adak, they found not a single carcass. What in hell, wondered Estes.

What finally convinced him of Tinker's sanity was Clam Lagoon. Clam Lagoon lay adjacent to Kuluk Bay, where the otters were vanishing in droves. Yet the counts at Clam Lagoon remained uncannily steady, straightlining merrily across the charts. The pivotal difference between the two could be traced to an act of Ice Age geology. Across the mouth of Clam Lagoon lay a natural breakwater of glacial rubble, creating a shallow channel all but land-locking the lagoon. Kuluk Bay, on the other hand, was wide open to the ocean. More to the point, the shallow entrance to Clam Lagoon was a barrier to killer whales.

In retrospect, it suddenly made all the sense in the world, and none at all: the sudden appearance of killer whales skulking among the kelp beds in the early nineties; the subsequent spate of eyewitness attacks—a rash of nine in a four-year span (with more on the way)—compared with a handful of vague references and speculations from the previous two centuries; the free-falling censuses of otters; the utter absence of bodies.

But it was one thing to finger killer whales for a few hundred deaths at Adak, and another to indict them for an ocean-wide massacre. Across the Aleutians there were by now some forty thousand missing otters to account for, and maybe several hundred killer whales as potential suspects. So few killer whales could never eat so many otters . . . could they?

Estes turned to a fellow professor at Santa Cruz named Terrie Williams, who, besides being his wife, was a marine mammal physiologist. Williams determined that it took somewhere between 160,000 and 240,000 calories each day to power a killer whale. She then had several dead sea otters ground to hamburger, burned in a bomb calorimeter, and measured for the heat generated. It turned out a single swallowed sea otter in the belly of whale would provide 40,000 to 60,000 calories, or about the same as a baby sea lion.

From there it was a simple exercise in math. On a hypothetically strict diet of sea otters, a killer whale would need on average four to five a day. At that rate, Williams concluded, every one of those forty thousand sea otters gone missing over the last six years could have been accounted for by the appetites of as few as 3.7 killer whales.

"Uh uh," replied Estes. "Go back and run that again."

Williams returned. "Three-point-seven killer whales," she confirmed.

The blurred picture of the Aleutians had finally refocused in Estes's mind, if now a much grayer image than the one he had thought he knew. Sea otters were disappearing en masse, kelp forests were being clear-cut by urchins. It was like stepping back to the bloody days of the promyshlennik. Only this time, the suspected demon was no swarthy seaman with a club, but a national icon, performing star of marine parks and Hollywood films, carrying stage names such as Willy and Shamu.

With due trepidation, Estes, along with Williams, Tinker, and Dan Doak, a fellow ecologist at the University of California at Santa Cruz, submitted their death-by-killer-whale hypothesis to the journal *Science*. And with its publication in October 1998, the sea otter once again shone in the public light, albeit this time partly obscured by the menacing, sickle-shaped shadow of a great black dorsal fin.

The Brewing Storm

The *Science* paper landed with a splash. Major dailies on both coasts ran headlines: ORCAS GOBBLING ALASKAN OTTERS. KILLER WHALES PUT ALASKA SEA OTTERS AT RISK. *Science* itself lauded the findings in its news of the week: "Experts call the study a vivid example of an ecological cascade operating on a vast scale." Among those experts was the godfather of ecological cascades, Robert T. Paine, who said: "It's a heroic effort, and it's a terrific find."

It was an unsettling find as well. If true, this cascade had spanned a sea. It had reached from open ocean to coastal kelp, from the apex of the oceanic food pyramid down to one of its most fertile foundations. Along the way, it had exposed a dark potential of the ocean's topmost predator when tipped just a little off kilter.

But why? Why—after centuries of presumably peaceful coexistence— might the killer whales have so suddenly taken to chasing sea otters? Why would a whale-chasing, seal-swallowing killer whale switch, as one ecologist put it, "from eating steaks to eating popcorn?"

It was a question that concerned more than sea otters. It turned out the missing otters and their aberrant killers were only the latest in a succession of ecological phenomena sweeping across the northern seas since the 1970s. By the time anyone had recognized the sea otters failing, harbor seals, northern fur seals, and Steller sea lions were already a decade or two on the skids. The Steller sea lion, a buff-colored behemoth weighing as much as a ton, had by 1997 fallen so drastically as to land on the federal endangered species list.

Conservationists and the mainstream of marine biologists had come to believe that the answer was fish. Over the decades, the waters of the North Pacific had undergone cyclic swings in temperature, sending tremors through the oceanic food chain, flipping and flopping the stocks of fish that fed the masses. And shifting climate or no, there for anyone who cared to view it floated that omnipresent fleet of factory trawlers, some longer than a football field, mining fish by the megaton from the Aleutian waters. By the twenty-first century, the Alaskan fishery was accounting for more than half the total weight of the entire country's landings, with an annual catch in groundfish worth $1.5 billion. The trawlers and fickle temperatures, acting alone or in tandem, had to many minds become the de facto culprits in the collapse of the northern seas' fish-eating seals.

At the time, Estes had no quarrel with the fishy logic for the pinniped collapses. What mattered was that their disappearance could explain his missing otters: For want of seals, reasoned Estes, the killer whales had taken to eating otters.

But soon the stock explanations began to unravel under a barrage of scrutiny. One of the consequences of listing the Steller sea lion as legally endangered was the legal threat of closing those fisheries suspected of endangering it. Which led to the sudden appearance of a bargeload of federal research money, which in turn spawned a welter of competing hypotheses as to whom or what to blame. Another was an order from Congress for the National Academy of Sciences to sort through those hypotheses. The academy's blue-ribbon panel was chaired by Robert T. Paine and included eleven other ecologists, zoologists, oceanographers, and marine biologists, James Estes among them. By the time they had finished, they had unveiled a new prime suspect in the sea lion's demise.

Paine and his committee just couldn't explain the sea mammals' free fall on account of poor nutrition. It was possible that toxins or plague were involved, though evidence was slim. Whatever was killing the sea lions, it was killing them quickly. "Bottom-up hypotheses invoking nutritional stress are unlikely to represent the primary threat to recovery," the panel determined. Or in simpler terms, "The question shifts away from 'Is it food?' to 'Were they food?'"

It was the sort of scenario that repeatedly brought to suspicion a more violent sort of death, as by a bullet. Or, say, a very large set of jaws.

It was no secret that despite legal protection (allowing for subsistence hunting), Steller sea lions were still being shot as a habit by Aleut islanders and hateful fisherman. As for other predators, the committee deemed the resident sharks as lacking in both sufficient abundances and seal-eating tendencies to qualify as a smoking gun. But there was a third candidate with the pedigree and the appetite to potentially supply all the firepower necessary. Killer whales were indeed common in sea lion waters, in numbers enough to at least warrant theoretical concerns. (Terrie Williams's analysis of killer whales requiring around 200,000 calories per day had made that possibility quite plausible.) Moreover, killer whales had a deep history as sea lion gourmands. A biological survey from 1887 reported that the Steller sea lions "have a dreaded enemy in the Killer Whale, which pursues and captures them at sea and about their rocky resorts. The native hunters when at sea frequently see them leaping high out of the water in useless endeavor to escape their pursuers. At such times they say it is dangerous for an umiak or other small boat to be in the vicinity, as the animal, in its terror, will sometimes leap into and wreck the boat."

For those preferring fresher evidence, in 1992 a killer whale had washed up dead on a beach of Prince William Sound, its stomach containing ID tags once attached to the flippers of fourteen Steller sea lions.

For Estes, who was still watching his otters vanishing down the line, these were intriguing and troubling developments. If the killer whales, and not the fishing boats, were the ultimate cause of the sea lions' decline, that reopened the question why. He turned to his colleague Alan Springer from the University of Alaska at Fairbanks for a fresh perspective. Springer's authority lay in the broadscale ecology of the Bering Sea; he was an expert on the ecosystem's flux and flow, from plankton blooms to seabird crashes and back. Springer was just far enough removed from the entrenched combatants in the sea lion wars to free himself for more irreverent hypothesizing. What if, Springer offered, those blaming the fishing boats were only half wrong? What if the boats that had started this whole mess were whaling boats?

Decades before factory fishing boomed upon the Bering Sea, another great invasion had visited with the end of World War II, when the industrial whaling fleets set sail. The new battleships were swift, diesel-powered cruisers armed with cannons firing exploding harpoons. Whalers from Japan and Russia, rushing to rebuild battered postwar economies with this sea-bound bonanza of meat and oil, pushed into the demilitarized waters of the Pacific, across the Aleutians to the North American coasts. They came chasing the biggest and fastest of whales, the blue and fin, sei and sperm whales, in a massive blitzkrieg of the world's biggest animals.

By the time the blitzkrieg went bust a mere twenty years later, half a million great whales had been mined from the North Pacific and southern Bering Sea. Taking into account just the troubled waters of the Aleutians and Gulf of Alaska, the harpooners had consumed three and a half billion pounds of whale. And that, to Springer's and Estes's growing suspicion, amounted to removing one towering mountain of killer whale sustenance. Maybe for want of whale meat, the killers had turned to smaller fare.

Weapon of Mass Destruction

In 2002, Estes took his and Springer's nascent theory for a road test. To an auditorium packed with predator biologists and conservationists in Monterey, California, he delivered the conference's plenary address, so innocently titled "Ecological Chain Reactions in Kelp Forests."

A tall handsome figure at the podium, squint-eyed and trim-jawed and soft-spoken in the classic Clint Eastwood tradition, Estes called to mind the master's flair for understated drama. "Do carnivores matter in this web of life?" began Estes, foreshadowing something other than a simple discourse on seaweed.

"Big carnivores occur in every ecosystem of the world, or at least they did at one time," Estes continued. "The question of whether they matter depends on the resolution of top-down versus bottom-up forces. If bottom-up functions to the exclusion of top-down, then top-down doesn't matter. But if top-down forcing is important, then the carnivores might matter a great deal. I'm talking about a ripple effect, rippling all the way to the base of the food web."

The story thus began with a recap of Estes's own historic discovery of the sea otter as keystone keeper of the kelp. Then on it went to the startling crash of the mid-'90s, the return of the urchin barrens, the utter bewilderment. "I initially just didn't believe it. My mind-set was that these populations should be increasing," he said.

Enter Tim Tinker and the killer whales, the otters' port in the storm at Clam Lagoon, the astounding caloric calculations of Terrie Williams—all pointing to the mighty potential of the ocean's apex predator.

"Why did all this happen? I'm less certain about what I'm going to tell you," Estes told the crowd.

For lack of sea lions, Estes explained, the killer whales might be switching prey. Which of course then raised the $1.5 billion question: Why the lack of sea lions?

"That's a tough one," said Estes. "There's a lot of disagreement. There's emotion, money, and politics. There are entrenched theories." But there was also, he revealed, the soon-to-be published conclusions of the National Academy of Sciences, which had taken a crowbar to those same entrenched theories. For all the respectable scenarios invoking starvation of the sea lions by fishing boats or climate change, it was at least as plausible that something had simply eaten them up.

So, continued Estes, here slowly unwrapping his bombshell, if killer whales had eaten all the sea otters, might not they have decimated the Steller sea lions as well? Considering their needs and capacities, as few as twenty-six killer whales could have done the job.

With note-takers in the audience now rechecking their pads—did he say, twenty-six?—Estes flashed an image on the big screen, bringing a collective gasp. It was a stark studio shot with black backdrop, the frame filled by a pair of enormous fangs. At a glance, they looked identical. One indeed belonged to a killer whale, confirmed Estes. The other had come from a *Tyrannosaurus rex*.

The killer whale came equipped for bigger prey than mere one-ton sea lions, said Estes. "Killer whales fed on all the great whales. Early whalers called them whale killers."

Early whalers were, of course, whale killers too. Their postwar massacres had left the competition desperate, Estes suggested. For want of larger prey, killer whales had begun picking off the next convenient prospects down the line, from harbor seals to Steller sea lions to sea otters, which had apparently collapsed in suspicious sequence across western Alaska.

It was an ecological chain reaction that had spanned sixty years and six links in the food chain, from World War II to the twenty-first century, commercial whalers to kelp forests. "If you were to go out there today, you would not have a clue to what happened," Estes told the carnivore conferees. "Without some notion of history you would have no inkling."

"Carnivorous animals are important," ended Estes. "We have to stop thinking of them as passengers on this earth and start thinking of them as drivers."

Shots Across the Bow
Later that year, Estes, Springer, and seven collaborators submitted a formal account of their whaling hypothesis to the journal *Science*. And sometime after that, their star treatment came to a strange and inglorious end.

To the authors' befuddlement, the *Science* paper came back twice—first with a few encouraging suggestions for revision, the second time with a fatal rejection, appended with a caustic dismissal from one of the reviewers: *Everybody knows that Steller sea lions starved to death. That's a fact.*

Everybody, to Springer's and Estes's understanding, included neither themselves nor the National Academy of Sciences, which had just published a book effectively capsizing the conventional wisdom on the sea lions' nutritional demise. Their panel's chairman, Bob Paine, also took issue with *Science's* premature dismissal of the whaling hypothesis. It appeared the old guard of the sea lion debate was still wielding considerable political power over the scientific community. So Paine then wielded his. As special contributor to the *Proceedings of the National Academy of Sciences*, a journal with prestige of its own, Paine himself forwarded Springer's whaling paper for publication.

The lines had been drawn. And soon some of the troops grew skittish. Just before Paine submitted the paper to the *Proceedings*, one of its authors withdrew his name. Douglas DeMaster, director of the federal Alaskan Fisheries Science Center, begged off, citing unsettled scientific disagreements with his coauthors and concerns of conflicting interests. (DeMaster was a member of the International Whaling Commission.) By whatever reason, DeMaster's untimely departure heralded the heat to come.

In October 2003 the *Proceedings* published Springer and company's paper, "Sequential Megafaunal Collapse in the North Pacific Ocean: An Ongoing Legacy of Industrial Whaling?" Its authors might as well have stoned a hornet's nest. A stinging barrage of e-mailed vitriol and in-your-face insults came shooting back: *You've let the fishermen off the hook. People are going to be shooting killer whales over this. This is scientifically irresponsible.*

To Springer's crew, it became loudly apparent that their well-meaning hypothesis of sequential megafaunal collapse had breached the bounds of cordial scientific discourse. Pet theories had apparently been trod upon. Egos had been bruised. "There are a lot of agendas out there in the research community and in people's professional lives," speculated Springer. "And you don't want to have the rug jerked out from under you by some upstarts who aren't killer whale people, suggesting Why didn't we think of that?"

Not all the angst, though, was necessarily in vain. Anyone looking to excuse the trawler fleets for ecological damages (of which there were many) now had a conspicuous new scapegoat in the killer whale. Within three months, U.S. senator Ted Stevens, an ally of the Alaskan fishing interests, had proposed that the government begin looking into this new evidence "that rogue packs of killer whales" were to blame for the North Pacific's endangered sea lion.

As for Alaskans taking up arms against killer whales, biologists who'd spent time among them insisted the threats were not hollow. "The community out there will use any excuse to harm these animals," said Marilyn Dahlheim, a biologist with the National Marine Fisheries Service, who for years had been following and photographing Alaska's killer whales. "I look at a lot of pictures of killer whales, and I see a lot of bullet wounds."

If to no better end, Springer and company's killer whale hypothesis had forced everybody concerned to take a fresh look at the animal they thought they already knew.

The Killer Whale

Orcinus orca is by far the largest and most lethal member of the dolphin family. They are a toothed whale growing to thirty feet long, readily identified by striking contrasts of white on black, and a sickle-shaped dorsal fin that on males may stand six feet tall. The head is massive and blunt, the tail a muscular fluke that can propel a six-ton body through the water at thirty miles per hour, arguably the fastest of all marine mammals.

The killer whale is the greatest predator on Earth, sea or land. It outweighs the largest land carnivore, the polar bear, by eight times, and occasionally eats them too. Killer whales are known to kill six-inch herring as well as sixty-foot blue whales. The killer whale attacks nearly everything that swims, including some that maybe shouldn't; they are on record as having eaten dog, deer, and a moose. Great white sharks have been seen dead in the jaws of killer whales, and live ones apparently flee from the mere sound of them. Scientists opening the stomachs of killer whales have found the remains of fellow dolphins and porpoises, fish, octopuses, sea turtles, great whales of various forms, the occasional seabird, strands of seaweed, rocks (nine-pounds' worth in the stomach of one killer whale), and recently of particular note, sea otters (five inside a killer whale found dead in Prince William Sound in 2003). The only dolphins yet to be documented as killer whale's prey is the family of river dolphins. The only species of seal known to be safe from killer whales live in lakes.

The killer whale by necessity eats enormous quantities. Its metabolic furnace burns at twice the rate of terrestrial predators, ten times that of the largest sharks. On average, the killer whale must eat up to 5 percent of its body weight each day, which for the biggest bull whales comes to six hundred pounds. One excavated stomach contained the remains of fourteen seals and thirteen porpoises.

The killer whale's teeth are all essentially the same, forty to fifty-two big, conical interlocking pegs, pointing slightly backward in the jaw. None are for chewing; all are for seizing and ripping. Their maw is capable of swallowing baby seals whole. What a killer whale can't swallow, it rips to manageable chunks, often with a frenzied shake of the head.

For its predatory style, the killer whale is often compared to the wolf. It hunts in familial packs, combining guile, teamwork, tenacious resolve, and stamina to subdue quicker or more powerful prey.

Its hunting techniques are as varied as styles of art. Killer whales in Puget Sound herd salmon with their bodies and sonic voices, gathering the fish to be eaten one by one—the killer whale's version of fishing in a barrel. Killers who chase herring in the Bering Sea have learned to gather the morsels by bashing a school of them with a slap of their tail, then picking the stunned ones like apples shaken from a tree.

Mammal- and bird-eating whales employ a range of tactics all their own. If they are after a penguin or seal huddled upon the ice, they may rocket from the depths to smash through the ice. In the frozen waters of Antarctica, packs of killers have been known to tip ice floes to spill stranded seals into waiting jaws.

A few killer whales in the Crozet Archipelago of the southern Indian Ocean, and half a world away on the South Pacific shores of Patagonia, have independently learned the same technique for snatching seal pups off the beach. They approach silently under cover of an incoming wave, barreling from the wall of water like a train out of a tunnel, then grabbing an unsuspecting seal pup or sea lion at water's edge and thrashing the little animal like a rag doll.

Killer whales innovate and copy success. A killer whale held captive at Marineland in Ontario, Canada, learned to fish for gulls. The innovator, a young male, would spit out a fish from his daily rations. He would sink and wait out of sight for a scavenging bird to alight on the floating bait, then lunge. His little brother soon picked up the trick, followed by their mother. Eventually the entire killer whale community of Marineland had taken up the hobby of gull fishing.

Killer whales have also learned to train people. During the mid-1800s, a certain savvy pod of killer whales took to alerting the whalers of Twofold Bay, Australia, whenever there were humpback or right whales passing by. The killer whales would raise a ruckus, the whalers would spring to their boats, and as a team the two would go hunting together like hound and houndsman. The whaling men would reward their accomplices with first dibs on the tongues and lips of their catch.

Such predatory crafts and capabilities were common knowledge to any serious student of killer whale biology. Yet when it came to considering Springer and Estes's interpretation of killer whales as whale killers, the heads started shaking no. The idea was too much to swallow. The logbook accounts of ancient mariners seemed too sketchy, too fantastic to make a case. Modern biologists who had spent hundreds of thousands of hours in the pilothouses and observation decks, without ever having witnessed an attack, could hardly believe they'd been missing the show.

Estes, on the other hand, was not at all surprised by that possibility. Estes, who had spent thirty years staring at one of the most accommodating subjects a wildlife biologist could hope for, had never seen a sea otter give birth. "Not only have I never seen them give birth, no one has ever seen them give birth," he said. "I've actually gone through exercises: What would I expect to see, given the number of hours out there, and how few births occurred, how many would I expect to see? The expected number is about one. And we've seen zero."

Ambush Alley

A funny thing happened during the killer whale debate. Observers began noticing killer whales killing other whales by the score.

In California's Monterey Bay, through which gray whales pass on their annual migration from calving grounds in the Sea of Cortez to summer feeding grounds in the Bering Sea, professional whale-watchers looked on as killer whales started attacking gray whale mothers and calves with intensifying frequency. Between the years 2003 and 2004, the number of witnessed attacks went from two to two dozen. Monterey Bay had become known to insiders as Ambush Alley.

The same thing began happening as the grays rounded the Alaska Peninsula, funneling north through False Pass into the Bering Sea. There again the killer packs were roaming, taking as many as one in three of every gray whale calf that swam the gauntlet.

Ambush Alley and False Pass seemed to have rather suddenly become busier places. Whether a phenomenon of mounting attacks or the heightened vigilance of marine biologists, there was no denying that a certain faction of killer whales were making an increasingly conspicuous habit of killing gray whales.

But it was still too hard for many to imagine killer whales attacking the greatest of the great whales—the fin, the blue, and the sperm whales, some of which were double the mass of the gray whale. Never mind that in the late 1970s, a research vessel off the tip of Baja had happened upon, and graphically filmed, a sixty-foot blue whale being eaten alive by a pack of thirty killer whales. That crew had witnessed in gory detail a five-hour assault that left the blue whale hideously mutilated, its tail shredded, its dorsal fin eaten, and a coffin-size crater of flesh and blubber excavated from its spine.

But if there was one whale in the ocean that might have seemed invulnerable to the ocean's largest predator, it would be Physeter macrocephalus, the sperm whale—legendary slayer of giant squid, monster hero of Moby-Dick. The sperm whale was a formidable target reaching sixty feet long and forty-five tons, eight times the bulk of a killer whale. Before October 21, 1997, maritime history had left only nine records of attacks by killer whales on sperm whales, none of them lethal. On that day, some eighty miles southwest of California's Morro Bay, whale biologists Robert Pitman and Susan Chivers and several researchers aboard the research vessel David Starr Jordan recorded the tenth attack.

Coming upon a curious commotion in the water, they found a pod of sperm whales huddled face-to-face in a ring of bodies radiating tail-out, against a gathering mob of killer whales. As the crew watched, waves of killers converged from all points in the sea, taking turns battering and mauling the ring of sperm whales. One of the sperm whales was hauled from the circle and savagely set upon. Two others broke from the ranks of the rosette to retrieve their comrade, nudging it back to the group. "We see this same heroic scenario several times," Pitman and Chivers later wrote. "We are aware of our own tangled emotions as we watch in horror and fascination."

By the time night fell, Pitman and Chivers had recorded one fatality for sure and could only speculate gravely about the survivors. One whale had been filleted alive, dragging "a yawning slab of blubber . . . . as trig as a mattress." Another had been conspicuously ripped to the ribs. One whale was dragging its intestines in the water, while yet another had its jaw broken to a right angle. As the whales were left to the darkness, it was hard not to conclude that this entire herd of sperm whales may have succumbed to injuries from this single ferocious assault.

So much about the attack suggested to Pitman and Chivers that this was no fluke encounter between two otherwise unfamiliar species. The reflexive marguerite formation of the sperm whales (a cetacean forerunner to "circling the wagons"), and the coordinated, sequenced attack of their killers, suggested an innate choreography rehearsed and honed over many generations of high-stakes testing between the two—no matter the minuscule number of witnesses. After considering the odds of seeing just one, let alone ten attacks in such an immense, unmanned theater as the open ocean, sperm whale expert Hal Whitehead later calculated a typical female sperm whale in her sixty-year lifetime suffering, on average, 150 attacks by killer whales.

Before witnessing the attack, Pitman and Chivers were among the majority who had assumed the killer whale all but incapable of bringing serious harm to such a beast as a grown sperm whale. Five hours later, they saw the ocean's apex predator in a reverent new light. In their popular account written for *Natural History* magazine, the two expressed their epiphany: "This image of gentle giant may be ingrained in many people's minds, but the name 'killer whale' is an appropriate reminder that this species consumes huge numbers of marine mammals annually and that its predatory habits are a significant force in shaping marine communities."

But to many minds, a few extra sightings and a lot of killer whale potential still did not add up to the sequential collapse of marine mammals as proscribed by Springer et al. And to those minds, it was their professional duty to publicly say so. Between 2005 and early 2007, a fleet of papers appeared in the marine science literature harpooning the sequential megafaunal collapse hypothesis, saving a few subtle barbs for its authors. The brigade of rebuttals culminated with a remarkable thirty-seven-page broadside in the journal *Marine Mammal Science*, written by marine biologist Paul R. Wade of the National Marine Mammal Laboratory in Seattle and a crew of no less than twenty-three coauthors, who concluded their critique with a dismissive sniff: "Hypotheses which tender neat explanations for apparent changes in trophic relationships are often wrong; in this regard, it is all too easy to confuse parsimony with oversimplification."

Landward

Springer and Estes returned fire to Wade and his crew, submitting a defiant little volley of diplomatic thank-you's and point-well-taken's, and by the way, we still believe and dutifully maintain that our argument remains a plausible one worthy of serious consideration.

Whatever one believed—whether an era of bygone whalers had or hadn't triggered such ecological mayhem nearly half a century earlier—either side could hardly be proven wrong. The trail of evidence was, in some cases, half a century too cold. And on that, both sides agreed. What stunned the Springer camp was not the disagreements over the murky facts of history, but the seeming allergic reaction to the notion that predation quite plausibly had sent a biological tsunami through an oceanic ecosystem.

"People don't like it because they don't want to think of killer whales as having a role in the ecosystem so potentially profound," Springer speculated. "It just doesn't jibe with their sense of nature."

"The thing people have forgotten is that the title of that article ends with a question mark," said Bob Paine, observing from ringside. "All the adversaries are forgetting that question mark."

Springer's name may have come first on the incendiary SMC paper, but it was the second author, Estes, who received the brunt of the backlash. He had been its main messenger from the start, and he had come away with bruises and a darkened view of his profession. One day in October 2006, nearly ten years into the battle that was never intended to be, Estes reflected with the tone of a wearied but unyielding warrior. "It could be we're both wrong. After all, I have spent a hell of a lot of time working on this. I'm humble enough that I have some respect for the other possibility.

"On the other hand," he added, "I'm arrogant enough to think that I'm not that naive. I've seen enough of this part of nature—the broad-ranging and important impacts of predators—to the point that I really believe in my heart that it is a recurrent and important phenomenon across much of nature. The part of nature I've seen most clearly has made me feel some confidence that this is an issue that's real, and the world needs to recognize that it's real despite the fact we don't know very much."

To be fair, Estes's view of nature had not exactly developed in a North Pacific vacuum. Nearly ten years before at a workshop near Tucson, he had met an ecologist named John Terborgh. Terborgh, along with conservation biologist Michael Soulé, had organized a gathering of thirty experts for a sky's-the-limit discussion of what it would take to save the wild nature of North America. Terborgh's session, to which Estes contributed, was "The Role of Top Carnivores in Regulating Terrestrial Ecosystems." After some twenty-five pages of anecdotes and experimental evidence marshaled from across the western hemisphere (citing, among others, Paine's starfish and Estes's otters as paragons of proof), the synthesis was summed by Terborgh and his conferees in decidedly strong terms: "Our current knowledge about the natural processes that maintain biodiversity suggests a crucial and irreplaceable regulatory role of top predators. The absence of top predators appears to lead inexorably to ecosystem simplification accompanied by a rush of extinctions."

One of the darker developments leading to Terborgh's bleak assessment had come from Terborgh himself, whose team had just begun sending dispatches from a series of forested islands in a Venezuelan lake. If Estes had harbored any doubts about his worldview of predators—schooled as it was on sea otters and killer whales—Terborgh's frightening news from his terrestrial laboratory would ultimately shatter them.

FIVE Ecological Meltdown

Chance favors the prepared mind.

—Louis Pasteur

IN 1986, NEAR THE confluence of the Orinoco and Caroní rivers in east-central Venezuela, the gates finished closing on a huge hydroelectric dam. The rising waters flooded a tract of tropical forest as broad as the state of Connecticut under a lake that came to be named Guri. Left poking above the surface of Lago Guri was an archipelago of hundreds of hilltop islands, ranging in size and quality from half-acre islets to sprawling 1,800-acre semiwildernesses. The disparities in size had created one critical distinction: All but the very biggest among them were too small for large predators. The birth of the Guri archipelago was the scientific windfall the ecologist John Terborgh had been waiting for.

By the time he caught wind of Lago Guri in 1990, Terborgh, had established himself as a premier tropical ecologist, a multifaceted master of the rainforest's unrivaled roster of life-forms. In the field, Terborgh could hold his own with the foremost botanists, ornithologists, and primatologists. He had published major treatises over an unheard-of breadth of tropical ecology, from the ecology of migrant songbirds to the sociology of neotropical monkeys. To travel a quarter mile on the trail with Terborgh was to spend five hours examining every insect, snake, frog, bird, flower, and tree along the path, addressing each by Latin name, occupation, and relation to all others.

Terborgh had a special ability for seeing all as a working whole, linking bottom dwellers to treetop specialists, big and fierce, small and meek. Among the inner circles of tropical ecology, he was "the guy who knows what's going on."

Terborgh was also fearless (opponents would say reckless) about reaching into his bottomless rucksack of natural history knowledge and pulling out big sweeping theories of ecology. Over the previous decade in particular, Terborgh had begun "harping on a theme"—a theme he had most baldly declared in a 1988 essay titled "The Big Things That Run the World." At the top of Terborgh's list of big things perched the big predators.

After spending a third of his adult life probing their place in the tropical forests, Terborgh was convinced of the predators' disproportionate and heretofore unheralded importance. "If what I suspect is true, the top predators in this system—jaguar, puma, and harpy eagle—hold the key to its stability and to the maintenance of its extraordinary diversity of plant and animals."

But Terborgh's suspicion wasn't to be answered by a few stubborn starfish on a point of rock. His subjects of concern were out there slinking like phantoms through great vastnesses of dark jungle, not to mention armed with tools capable of dismembering intrusive field biologists. Long-lived and wide-ranging, the superpredators' impacts were to be measured on the scale of years and decades, not the months of a master's project. Terborgh needed a place where the big predators could be systematically removed, the ecological consequences rigorously gauged. He needed something on the order of a miracle. But as Pasteur once declared, "Chance favors the prepared mind," and when the dam created the archipelago of Lago Guri, John Terborgh was more than ready. In truth he'd been training for the moment from the time he was a toddler.

The Boy Naturalist

John Terborgh was born in Washington, D.C., in 1936, the eldest of three children who grew up across the Potomac River in what was then the rural outskirts of Arlington, Virginia. His home was a small white house built on an abandoned pasture at the end of a country lane. Young John played no ball; there were no boys his age. There was instead a forested stream that ran more than a mile from his backyard to the palisades of the Potomac, crossing but one street along the way. By the time he was five, John had taken to disappearing into the greenery of Donaldson Run, roaming wood and stream, turning up snakes and turtles and salamanders, many of which would eventually end up in the home menagerie. The Terborgh home became one of snake cages on the porch, turtles in the bathtub, flying squirrels in the bedroom. For a time the Terborgh family also had a cleaning lady, whose visits abruptly ended the day she met one of John's serpentine specimens slithering down the hall.

Nights, Terborgh's mother would read to him tales of wild places and exotic creatures, with John imagining himself in the living cathedrals of James Fenimore Cooper's American forest primeval or tracking pythons and crocodiles through the jungled tropics with herpetologist Raymond Ditmars. Weekends the Terborgh kids followed in the footsteps of Uncle John Murray, who led the family on field trips to Shenandoah National Park. The kids would follow behind like a brood of ducklings, poking and putzing and rummaging in Murray's wake, hailing him with questions: What's this flower, what's that tree, whose bird song is that? In the shadow of Uncle Murray, John developed his signature style, a meticulous, all-things-considered exploration of nature, from the tiniest to the grandest forms of life.

Terborgh burned his way through the field guides, exhausting the inventories of his backyard Mid-Atlantic flora and bestiary, building an encyclopedic grasp of all things nature. By his teens, he was bird-watching under the wing of Chandler Robbins, one of the country's leading ornithologists and author of the *Golden Guide to Birds*, one of the most popular field guides to birds in U.S. history.

As sure as the arrival of warblers in May, the post–World War II baby boom and its bulldozers inevitably came to Terborgh's end of the road and rapidly squeezed his seemingly boundless backyard wilderness to a green ribbon between suburban housing lots. Along with the forests, they took away both playground and playmates of the young naturalist. Terborgh fled for college and yet untrammeled places in the tropics, bringing with him memories of flowers and fauna at their richest, images that would serve as yardsticks against the coming changes. It was a habit Terborgh would find himself practicing all too many times in his career.

In between professorial duties as biology instructor at the University of Maryland, John began making good on his boyhood fantasies, trading his romps in Donaldson Run with excursions to the tropics. In the foothills of the Peruvian Andes, Terborgh came upon a valley he later declared as "the most beautiful place I had ever seen"—a valley that was soon destroyed in reminiscent fashion of his Arlington environs. "By 1972, the year I last saw the Apurímac valley, the population of colonos had swollen to more than a hundred thousand and was still growing steadily. By then, hardly a tree remained of the magnificent forest that had so recently filled the valley bottom," he wrote.

Over the next twenty years Terborgh explored some eighty forests in fifteen countries, visiting tropical wildernesses in every state of ecological repair. Along the way, he developed an eye for distinguishing the pristine from the pathological. He saw through the façades of idyllic green forests that unbeknownst to the inexperienced observer had been stripped of their distinguishing fauna.

He began imploring others to see this, too. As a professor at Princeton University, he would gather a hardy handful of ecology students during the winter semester break, pick a tropical spot on the map, and go explore whatever avifauna might await. The Princeton ornithology field camps, and their leader, became legend. "John was famous for having a place and a map and just kind of winging it as we went," said Scott Robinson, a former Terborgh student and now a leading authority on migratory birds of the New World tropics. "The beauty of it was that he did it without any fancy stuff. He doesn't need money, doesn't need massive planning. He can make do with a machete, wood, and a tent."

Terborgh and the students would spread invisible nets across gaps in the woods and take inventory of the forest's avifauna. They would come back with enough data for technical papers. But what the students valued most were those rare glimpses of the tropical forest as seen through Terborgh's eyes.

Said Robinson, "The thing that always boggled my mind when we would walk through the forest, he would always be grabbing at a plant, or looking at a flower, taking his glasses off, staring at it, pondering it, identifying it, looking at context—looking out from extreme structural detail all the way up to the structure of the canopy. You could see his eyes were always focusing up close and then looking at a distance up into the trees. Just seeing things. You could just tell he was seeing things we weren't seeing. I've known many of the world's great naturalists, but John is the finest I've ever known."

"The thing that struck me was the sort of breadth of his natural history knowledge," added David Wilcove, now a decorated conservation biologist who has since inherited his professor's old post at Princeton and takes students of his own to the tropics. "It was his ability to see patterns where the rest of us didn't. We might gather around the campfire in the evening and he would say, 'Did anyone notice that the bird community here is largely composed of nectar-eaters? Or that this area is very different from what you'd see in a lowland forest of South America?' He would, say, observe a dot-winged antwren foraging in eastern Mexico, and he would remark that it occupied a much wider range of heights in the canopy here than one you saw in Peru, where there were many other competitors and therefore had a much narrower foraging range. That was the kind of amazing thing he could do."

Barro Colorado

In 1970, Terborgh's tropical peregrinations brought him to a bizarre, 3,800-acre island in Panama. Barro Colorado Island had once been the dominant hilltop in a mainland forest, until in 1913 the construction of the Panama Canal left it standing in the midst of the new Lake Gatun. Barro Colorado, not long after its creation, had been blessed with protection as a biological reserve and laboratory, run by the Smithsonian Institution. It should have been an idyllic tropical wilderness, but even as Terborgh was arriving, just over fifty years after its formation, the island was already showing aging signs of insidious decay.

Terborgh on his first visit tagged along with the island's resident ornithologist, Edwin Willis, who had picked up where a line of earlier ornithologists had left off. Willis had been following birds on BCI for nearly a decade. Willis had discovered that of the 209 species counted in the earliest censuses of BCI, 45 were no longer to be found on the island.

Some of the disappearances were to have been expected. About half the island had once been slashed and burned by peasant farmers; as the forests grew back, those birds best adapted to field and pasture quite naturally fell out. But among the missing were also some eighteen forest dwellers, which, if anything, might have been expected to flourish in the returning forests of Barro Colorado. A few of them Willis had watched intently, year after year, as their numbers withered to single digits, then to single birds, then ... *poof*.

There was little in the way of science to explain it. And that became Terborgh's challenge. "As concerned citizens and biologists we are anxious to understand how natural diversity can be maintained in a world of rapidly diminishing resources," he began a 1974 article in the journal *BioScience*. It was by then becoming apparent to Terborgh in his wanderings—as well to many of his colleagues in their study sites around the world—that the diversity of living species was falling at a rate comparable to the greatest mass extinctions on record. A puzzle that had once fallen to the task of paleontologists, unearthing the die-offs from fossil beds and boneyards, had become the dilemma of modern biologists watching their subjects vanish before their eyes. Terborgh's eyes were as sharp as anybody's, and still those eighteen missing birds from Barro Colorado baffled him.

In general, the missing birds were either the larger species of their particular guild or those that typically foraged or nested on the ground. Of the eighteen, sixteen were still to be found on the mainland, which was in some places only a little more than a quarter mile away. He could imagine that the larger species of birds, with their greater demands for food and space, and their typically low numbers, would naturally be more prone to extinction. But the missing ground dwellers stumped him. "Why did that particular set of birds go extinct?" he asked. "They were just some garden variety songbirds. It seemed there was no particular thread that made them vulnerable." In concluding, Terborgh all but threw up his hands. "These are nontraumatic population declines compelled by unrelenting forces that are yet to be identified." That was as far as Terborgh could go in 1974. But for a long time afterward, that particular cast of missing birds on Barro Colorado haunted his mind.

Six years later, Terborgh was back with another stab at the answer. In the book *Conservation Biology*, a seminal volume announcing the formal emergence of a discipline spurred by the biodiversity crisis, he and Blair Winter contributed a chapter titled "Some Causes of Extinction." They noted two essential kinds of extinction: primary and secondary. Primary extinctions arose from such forces as fragmentation, the chopping of big populations into small and dangerously isolated populations, with their inherent susceptibility to freakish and fatal accidents. Primary extinction described the more obvious assault on the diversity of life, the fallout from the ubiquitous human saw as it cut through the last broad bastions of wilderness. But it was the more subtle and unheralded phenomenon of secondary extinctions that most intrigued Terborgh and Winter.

These, they noted, were of the kind that Robert T. Paine had famously demonstrated when he tossed the predatory starfish *Pisaster* from the splash zone of Mukkaw Bay, and watched half the species of its community disappear shortly thereafter. "Herein lies a potentially important area of research that has scarcely been breached," wrote Terborgh and Winter. "We know next to nothing about what consequences follow the loss of top predators in terrestrial ecosystems."

This was their cue to introduce Barro Colorado Island, in all its splendidly disguised decay. "Completely protected for more than 50 years, Barro Colorado has become a veritable zoo without cages," wrote Terborgh and Winter. "Medium and large mammals are unwary and remarkably abundant. These include the collared peccary, agouti, coati mundi, three-toed and two-toed sloths, tamandua, armadillo and howler monkey." The prolific, carefree denizens of Barro Colorado were of the same genetic stock as their scarce and skittish counterparts on the mainland. Why this menagerie remained so remarkably abundant and unwary, Terborgh proposed, was at least partly due to an unnatural freedom from being chased and eaten. Barro Colorado, at just seventeen square kilometers, was too small to support jaguars or pumas or the massive harpy eagle, the topmost predators of the tropical forest. (All three of them good examples of primary extinction via fragmentation.)

It was only conjecture, but it came well educated. Terborgh's tropical canvassings had calibrated his eye. No other place, not even those reserves banning hunting, approached Barro Colorado for the tameness and profusion of its mammals. What Terborgh saw in Barro Colorado was something other than a peaceable kingdom of tropical splendor.

Cases in point were those missing birds from the forests of Barro Colorado Island. Of the fifteen to eighteen birds that had vanished, many shared a certain facet of life history. Many nested on or near the ground. And of the suite of mammals whose populations had conversely soared since the island was born, two of them, the coati and peccary (southern cousins to the raccoon and pig, respectively), were notorious gourmands of eggs and nestlings. Along with the ground-nesting bird's typical and sundry difficulties of raising young in such a competitive environment, the added weight of an abnormal load of predators could easily drive the yearly production into the red zone. "All this is merely speculation," noted the authors, "but it has a persuasive ring to it."

One year later, two graduate students from the University of Illinois set out a line of wicker bird nests on the ground and shrubs of both Barro Colorado and the adjacent Panamanian mainland. They lined their artificial nests with leaves and placed a pair of domestic quail eggs in each. When they checked a day or two later, 6 percent of the ground nests on the mainland were missing eggs. On Barro Colorado, the losses tallied 88 percent.

Manu

Shortly after fleeing the Apurímac Valley in 1973, as his paradise came crashing down under the flood of *colonos* and drug lords, Terborgh came upon a far grander prize in Peru's Manu National Park. Nearly six thousand square miles of Amazonian lowland jungle extending unbroken to the thirteen-thousand-foot heights of the Andes, Manu circumscribed one of the wildest places left on earth. Manu was an embarrassment of biological riches, harboring nearly half as many species of mammals as all of North America and nearly twice as many species of birds. In a single acre of Manu, one could count a hundred species of trees. It was likely that Manu housed more species of life than any other park in the world.

The nearest semblance of civilization at Manu was a rustic biological research station called Cocha Cashu, which Terborgh promptly adopted as his tropical home and laboratory, becoming its chief scientist and administrator when not teaching in the States. The accommodations of Cocha Cashu included a desk, a kitchen with two kerosene stoves, a lake to bathe in, and a choice of several million acres in which to pitch your tent. It took three days' travel in a motorized dugout canoe to reach the inner sanctum. For the first seven years, there was no phone, no two-way radio, no outside communications whatsoever. Nobody ever just happened upon Cocha Cashu; the only primate visitors were the monkeys that gamboled before the open windows of the research station. It was John Terborgh's private villa on the Seine.

What mattered most for Terborgh's purposes was the fact that Manu, in its relative purity and wholeness, was a baseline against which to measure all other tropical forests of its kind. Manu was fully stocked with jaguars, the third largest cat in the world; and the puma, slightly smaller but more versatile, its species spanning latitudes from Patagonia to the Yukon, making game of everything from bite-size rodents to quarter-ton elk. Lording over all life in the trees was the harpy eagle, the world's most powerful and impressive raptor, an eleven-pound, monkey-seeking missile with warrior plumes and the talons of a tiger. Manu still harbored these superpredators of the New World tropical forests. But Barro Colorado, save for their sporadic visits from the mainland, no longer did. And that got Terborgh again pondering the maddening disappearance of Barro Colorado's birds.

In the early 1980s, mammalogist Louise Emmons of the Smithsonian Institution conducted a study of Manu's pumas and jaguars, as well as a smaller spotted cat called the ocelot. Emmons endlessly walked the trails, tracking the big cats by radio signals, collecting their scat, analyzing their diets. She tracked them day and night, and they, at times, tracked her, the steady beeping of the jaguar's collar following Emmons's footsteps through the dark.

Emmons also censused the jungle for the cats' potential prey. She took note of agoutis and pacas, two stout, seed-eating rodents of the forest; and capybaras, semiaquatic, nearly sheep-size rodents of the riverbanks. She tallied the piglike peccaries and the raccoonish coatimundis. And then she noted how often each got eaten.

Emmons discovered the predators sampling prey in proportions almost identical to their densities in Manu—lots of agoutis crossing the path meant lots of agoutis getting eaten. This surprisingly random mode of hunting meant to Terborgh the cats might be capable of regulating their prey. And by extension, it suggested that in places where the cats no longer hunted, those prey were likely to boom.

And indeed, the differences between Manu and Barro Colorado were astounding. The Barro Colorado agoutis outnumbered those on Manu by twenty times, the coatis by twenty-five times. As avid seekers of eggs and nestlings, the hyperabundant coatis became prime suspects in Terborgh's search for Barro Colorado's missing birds. While not suggesting the coatis had single-handedly exterminated the island's missing avifauna, Terborgh found it hard to imagine such a talented nest raider, in such abnormal densities, as anything less than an accomplice in the bird murders of Barro Colorado.

Over the years, challenges to Terborgh's reconstruction of Barro Colorado's demise would be many. Some questioned the accused coati's role, after finding no greater rates of nest predation on Barro Colorado than on the adjacent mainland. Others maintained that the little animals on Barro Colorado—extinctions notwithstanding—were doing just fine without big predators, thank you.

As stubbornly as Terborgh held his ground, he was well aware of that conspicuous lot of if's and but's in his argument. He realized the enormous and tenuous reach between Manu and Barro Colorado, two vastly different paradigms of wilderness, hemispheres apart. In his "Big Things" essay of 1988, in which he had openly lauded the powers of the superpredators, he had also openly opined for the "major research program . . . not yet undertaken."

Terborgh dreamed of a new laboratory—a fresh system of detached fragments, along the lines of Barro Colorado but with more islands and simpler faunas to sample. And he needed to be there from the start.

Lago Guri

In the spring of 1990, the primatologist Warren Kinzey mentioned to his friend John Terborgh that he was heading down to Venezuela to his new research site. The monkeys of Kinzey's interest had been stranded on some islands in the midst of a huge hydroelectric impoundment that had sprung up just a few years before.

"I'm going down with you," blurted Terborgh.

Terborgh spent two weeks on the giant lake with its nascent archipelago, growing giddy with its possibilities. The water that had risen behind the Raul Leoni Dam created a lake fifty miles wide and five hundred feet deep over what had once been the valley of the Río Caroní. Only the highest peaks of the hilly countryside had been left dry, surrounded by cemeteries of treetop tombstones. The islands came in all sizes, from one quarter acre to one and a half square miles, each harboring its own community of castaways. Most critical to Terborgh, many of the islands were decidedly too small to house the superpredators that he suspected ran the world.

Stepping ashore for the first time on the little islands of Guri, Terborgh was struck by their extreme asymmetry and exaggeration, flip-flopping between obliterating barrenness and explosive abundances. The islands presented a patchwork of incongruities, marrying the work of phantom bulldozers to some satanic gardener of hell's thickets. It was, in Terborgh's own words, "a god-forsaken place."

Terborgh returned to his teaching job at Duke University in a fervent state of anticipation. And for the next two years . . . he waited. And he steamed, as one grant proposal after the next was rejected: Too ambitious, too speculative, too long a wait for returns, came the replies. Terborgh bristled at their meekness—How skeptical, how reluctant, how cautious, he thought. In

1992, fortune stepped in, rewarding Terborgh with the coveted MacArthur Fellowship (popularly known as the "genius grant"), which provided him with a big sum of discretionary cash for his intellectual pursuits. Terborgh immediately poured the money into his Guri dream project. He assembled a team of scientists from across the continents. And the very next field season, one of the great experiments in tropical fragmentation was launched.

Beginning in 1993, and for two months every year thereafter, the field crews—Terborgh's students from Duke, local biologists, colleagues from institutions as far away as India—would descend upon the giant lake. From the mainland they would boat hours across the water to pinpricks of land, erecting makeshift camps in the Terborgh tradition, sheltering in tents, eating daily rations of rice or beans cooked on a gas stove, and bathing in the tepid waters of Lago Guri. The crews censused animals of all stripes—ants, butterflies, tortoises, monkeys, birds. They also surveyed the plants, counting trees and seedlings, following the forests' succession of life through the maturing of Lago Guri's newborn archipelago.

Guri's lack of living space had simplified the surveyors' task. Three quarters of the animal species found on the mainland were already missing from the islands by the time the crews arrived, most of them short-lived creatures unable to maintain themselves in such limited confines. As the islands got smaller, the casts naturally got shorter. What remained were skeleton crews of creatures surviving, in some cases, to numerical extremes.

They found trees crawling with *Cebus olivaceus*, the olive capuchin monkey, an artful, agile omnivore foraging from canopy to forest floor. And in time, those same trees began losing species of birds. On one island, during a period in which the capuchin population doubled, more than half of the avian species was lost. The Guri crew put out artificial bird nests stocked with quail eggs only to have every egg disappear. It was of no marginal concern that many of these disappearing birds had previously been employed as pollinators and seed dispersers of the Guri forests.

But among the primates, it was *Alouatta seniculus*, the red howler monkey, that came to most personally symbolize the depravity of Guri's bottom-heavy world. The Guri researchers would step onto an island the size of a suburban lawn, peer up, and see five howler monkeys staring down from a naked tree. Those five amounted to twenty to thirty times the typical densities on the mainland. But in order to appreciate the strangeness of that arrangement, it helps to first consider the brief textbook résumé of a mainland howler monkey.

Alouatta seniculus is one of a suite of six species of howler monkeys found throughout New World tropical forests from southern Mexico to northern Argentina. Prehensile-tailed primates typically living in bands of three to seven related individuals, howler monkeys are led by a large male, a husky, throaty monkey whose territorial roars can travel three miles through the jungle, letting neighboring troops of howlers know to stay away, and otherwise giving neophyte tourists the unnerving impression of lions in their midst.

The howler monkeys of Guri had been put under intensive watch by Terborgh's researcher Gabriela Orihuela. Nothing Orihuela had ever seen or heard of monkey society on the mainland could prepare her for the sights awaiting on Lago Guri's littlest islands. Free of fearsome predators—and hounded by hunger—the howlers of Lago Guri no longer lived in coherent groups but instead they slept in separate trees. For lack of contact, they seldom groomed. Those that did come together sometimes fought, inflicting savage wounds. Their babies never played. The monkeys of Guri grew thin. There were ominous signs of infanticide. The howler monkeys of Lago Guri no longer howled.

The commonly held and unscientific impression among Orihuela and the rest of the Guri crew was that the howlers of Guri simply didn't like each other anymore. In this supposed paradise free of predators, the group-hugging howler monkey had been sentenced to a solitary confinement in hell.

Even the favorite trees of the Guri howlers began biting back. Denuded by the relentless browsing, the trees started sending forth leaves increasingly spiked with bitter, nauseating toxins. Breakfasts for the monkeys became rituals of self-administered poisonings, with bouts of hapless gorging followed by episodes of perfunctory vomiting. Ostensibly freed from the top-down control of predators, the howler monkeys had entered a far nastier bottom-up realm run by plants.

Atta

Among the more striking features of the distorted Guri islands were the forests themselves. Denuded trees and tangles of vines and flesh-shredding thickets of thorns—these were ecological aberrations for which the howler monkeys could not be ultimately blamed. Something other than the monkeys had been eating the forest to such severity. In the predators' absence, the title of the most fearsome animal on the Guri islands had been passed to an ant.

The variety of concern was a leaf-cutter ant of the genus *Atta*. Dedicated farmers of fungus, *Atta* ants stream into the trees each day and return with mandibles grasping leaves they have harvested. Descending into their subterranean nests, the leaf-cutters chew the leaves to a pulp, preparing the compost upon which their sustaining gardens of fungus grow. In more typical forests of the tropics, the work of the *Atta* ants often goes unnoticed, their harvestings meager, their forays conducted under cover of night. On Guri, *Atta* society had taken a turn.

The Guri forests had left too little room for the chief predators of the leaf-cutter ants. Army ants—those fearsome Mongol hordes of the tropical forest floor that consume fauna like a ground fire—were absent in the inadequate confines of Guri. Another notorious plunderer of the leaf-cutter, the armadillo, with its tremendous foreclaws excavating like backhoes, went missing from all but the biggest of the Guri islands. A lethal family of parasitic flies, practiced at laying their eggs in the leaf-eater's heads, like so many others tended to be scarce on Guri.

The ants' predator-free life on Guri thus took on a frenzied pace. The *Atta* colonies spread like a rash, to densities a hundred times beyond the mainland norms. On the tiniest islands they merrily streamed about in broad daylight, undeterred by any pesky head-piercing flies, unchecked by the pillagings of army ants or the excavations of the armadillos. The rash as metaphor conveyed as well to the wreckage *Atta* wrought upon the Guri forests. The leaf-cutters' leafy trees gave way to thorny vines and walls of climbing lianas, the bane of the Guri researchers. The barren ground ran bright red with the soil of the ants' excavations. Unlike the howler monkeys, who at least recycled some of their nutrients by way of dung, the ants sequestered the greenery of the forest in their underground gardens, beneath even the reaches of the roots. While otherwise devouring the trees from above, they were starving them from below.

Post-Atta

The result was what came to be unfondly known by the Guri researchers as the post-*Atta* stage. The forest post-*Atta* was a depressing place where the leafiest, tastiest trees had been progressively replaced by walls of defensive vines and thickets of the thorniest, least-palatable plants of the forest, a place even the birds had abandoned. "It is one of the nastiest places you can imagine," said Ken Feeley, one of Terborgh's students and chief investigators on Guri. Feeley shuddered to remember the nastiest post-*Atta* island of all, an island the crew came to name Sudor, Spanish for "sweat." "We would go in with machetes, and we wouldn't know which way to turn," recalled Feeley. "There were more thorns than I ever thought possible. They're terrible places, the post-*Atta* forests. I don't know how else to describe them."

Feeley later came to learn that his predecessors had given Sudor a nickname of their own. They had called the island Difícil, Spanish for "difficult."

As the lianas gained control, they spread atop the canopies, blocking the light, smothering the trees. The seedlings and saplings of the Guri forests were dying faster than they could be replaced. No more monkeys, no more ants, no more trees. "The end point of this process," Terborgh and his colleagues would later write, "is a nearly treeless island buried under an impenetrable tangle of liana stems."

Said Feeley, "I wish everyone could go to Guri. Because I think our message would be crystal clear then, if they were actually able to see this place."

In the contorted kingdom of Lago Guri, Terborgh and crew had come to witness the green world hypothesis of Hairston, Smith, and Slobodkin turned upside down. "If you had any doubts about top-down regulation before," said Terborgh, "all you have to do is see a system regulated from the bottom-up. Once you've seen it, you realize you've never seen anything like this before."

Brown New World

In 2001, Terborgh and ten colleagues published the most famous of their dispatches from Guri, a *Science* paper titled, with Terborghian flare, "Ecological Meltdown in Predator-Free Forest Fragments."

"These observations are warnings," wrote Terborgh and his coauthors, "because the large predators that impose top-down regulation have been extirpated from most of the continental United States and indeed, much of Earth's terrestrial realm."

Under Terborgh's meltdown scenario, the grim cascade that had dismantled the little islands of Guri—the initial free-for-all of the herbivores, the collapse of howler monkey society, the pillaging of forests by the ants, the smothering shrouds of thorny vines—was the green world gone brown.

Look around, Terborgh implored. Look at the overgrazed grasslands of the arid American West, giving way to prickly scrub and thorn; look at the forests of Malaysia disappearing under assault by wild pigs, or those of the eastern United States being swallowed by white-tailed deer.

If Lago Guri was any indication, the inverse of a world ordinarily kept green by predators was not a pretty place to ponder. And one of those places, dearest to heart, happened to be Terborgh's own boyhood backyard.

SIX Bambi's Revenge

IN 2003, NEARLY A decade into documenting Guri's decay, Terborgh's study came to an inglorious end. A prolonged Venezuelan drought had drawn down the lake waters, exposing land bridges that turned the archipelago into a muddy amorphous mass. The miserable howler monkeys fled their prisons. The great experiment was over. But not before its images of ecological meltdown had been stamped into the conscience of conservation biology.

A year later, while taking a break from business meetings in Washington, D.C., Terborgh ventured a sentimental journey to the forested banks of the Potomac, to the baptismal parks of the boy naturalist. The sixty-nine-year-old Terborgh came away crestfallen. The wildflower garden of his youth had become a wasteland of foreign weeds. A forest floor once colored with bluebell, phlox, and spring beauty had been swept by an invasion of garlic mustard and Japanese stilt grass.

Thousands of miles and an ecological universe apart from Venezuela, Terborgh could not escape the specter of Lago Guri. "It looks like a war going on out there," he said. "That big tract of forest was eaten down to the ground. The whole character of it—it's just a ghost of what it was. And it's all deer."

Planet of the Deer

Terborgh had not just happened upon a rough spot in the woods, some bad stretch of scenery. He could have been dropped into almost any green space of the Capital region and come away with the same verdict. He could have looked into Rock Creek Park, the majestic forested corridor running so incongruously jade through the heart of D.C., and found the forest bereft of seedlings and overrun with unpalatable weeds. Mason Neck National Wildlife Refuge to the south and Bull Run and Riverbend parks to the west were besieged by deer herds as many as forty times thicker than what their attending biologists believed the forests could sustain.

Terborgh could have kept going, to the national parks of Catoctin in Maryland, to the Shenandoah of Virginia, the Great Smoky Mountains of Tennessee, all the way to Rocky Mountain National Park of Colorado, and found unhindered herds of deer and elk harvesting future generations of forest. He could have visited researchers in the field, somberly linking the burgeoning herds to disappearing life-forms, orchids to insects, red cedars to black bears. He could have crossed the Atlantic or the Pacific, northern hemisphere or southern, western Europe, Asia, New Zealand, Japan, to find the deer booming and woodlands withering. The family of deer—white-tailed or black-tailed, Sitka deer, fallow deer, roe deer, mule deer, elk or moose—had come to dominate food chains across the temperate latitudes.

It did not take an ecologist to spot the symptoms. Epidemiologists were pointing to North America's suburban epidemic of Lyme disease, a neurological infliction carried by a blood-sipping arthropod hosted by deer. Vehicular crash statisticians had been morbidly charting the rising tallies of deer flying through windshields and death on the highways. Cultural historians overseeing the battlefields of Gettysburg, Antietam, and Manassas were losing their battle to keep fields and forests of the Civil War's landscape from being eaten beyond recognition. And practically any gardener within foraging distance of the suburban deer could tell you that their war was already lost.

Terborgh's "all deer" hyperbole wasn't too far off. The mainland had, in effect, been Guri-ized, portioned into bite-size pieces, barred of big predators, and groomed for a takeover by the generalist masses. For the white-tailed deer in particular—given the hard times the species had endured in its evolutionary past—the twenty-first century had dawned as one immense rose garden.

Making of the Monster

The Pliocene epoch, in which the white-tailed deer arose three to four million years ago, was a tumultuous period of global cooling and drying rocked by pivotal spurts of tectonics. It came with the uplifting of ranges—the Cascades, Rockies, and Sierra Nevada—wringing moisture from the westerly winds, drying the plains in their lee, and cultivating the steppes and savannas on which the deer and most of the continent's modern hoofed animals evolved.

Two million years of Pleistocene ice ages followed, along with their emblematic mammoths and mastodons, giant ground sloths and saber-toothed cats. The Pleistocene was a time of mammalian giants, with five times as many kinds of large herbivores as today and twice as many forms of big predators chasing them. And through that gauntlet of Goliaths slipped a wispy, twig-nibbling deer.

Odocoileus virginianus, the white-tailed deer, was an opportunistic, weedy sort of creature, surviving, as one evolutionary biologist has put it, "between the cracks" of more specialized species. It assumed the role of the consummate generalist, a cosmopolitan herbivore of the grasses, buds, flowers, spores, and fruits, sampling widely across the kingdoms of plant and fungi. A deer well fed and unfettered by predators was an inherently fecund animal, capable of producing a fawn in its first year of life, and twins, triplets, and even quadruplets in its prime. On good range, a whitetail herd could double itself in two years.

Prospering in the Pleistocene commonly entailed escaping. The winter hide of the deer was the tawny gray of tree bark, its summer-slim profile melting through green walls of brush. Ever-twitching ears and vigilant eyes were backed by athletic getaway skills. On long thin limbs, levered all the way to its toenail hooves, the whitetail could spring away at forty miles per hour, leaping eight-foot shrubs and treefalls in the way. The deer guru Valerius Geist has watched whitetails turn and outrun his rocket-propelled capture nets fired from forty feet, and nonchalantly dodge arrows traveling seventy yards a second.

When hiding and fleeing failed, the cornered deer was equipped to fight, striking with sharp hooves and horns. Deer have been known to kill wolves, and cougars have been found dead with vital organs punctured by antlers.

Beyond surviving the Pleistocene, the white-tailed deer exploited it. The species spread from the Arctic Circle to sub-equatorial Peru. It spun off a new line of deer along the way, an open-ground version suited to the arid West, named *Odocoileus hemionus*, the mule deer. By the time Europeans arrived in North America five hundred years ago, some twenty-four to thirty-three million deer populated the continent (an admittedly tenuous estimate offered by the historians Thomas McCabe and Richard McCabe). And that is when the trajectories went haywire.

The Fall and Rise

Over the next three centuries, with settlers and Indians slaughtering deer for hides and trade, the deer numbers dipped. After which they plummeted. In the latter half of the 1800s—in the same half-century spree of market hunting and mindless gunning that brought the American bison from tens of millions to a few dozen individuals; that brought the passenger pigeon from billions to extinction—the whitetails of North America were eradicated over vast reaches of the expanding American empire. One Virginian, among many examples, took credit for killing more than 2,700 whitetails himself. McCabe and McCabe characterized the era's carnage as the greatest hunting pressure on wildlife *ever*.

Which was followed by one of the greatest wildlife *comebacks* ever. Market hunting was outlawed. States that had exterminated their deer began restocking, and pampering their returning natives. Foresters, farmers, and suburban architects cut swaths through the deep forest, laying open a landscape of fresh forage. Professional predator-killers vanquished the deer's enemies. By the mid-1900s the only wolves remaining south of Canada were rumored bands of renegades. Cougars in the East had been driven from all but a tiny pocket in the Florida cypress swamps, while those in the West harried to the wildest of hideouts. Even the human predator had been hamstrung—by choice. Deer seasons were shortened, bag limits were reduced, does were declared off-limits. The deer's American buffet was open 24-7, no predators allowed.

By the time Terborgh revisited his boyhood home at the start of the twenty-first century, the deer that he recalled hardly ever seeing in all his younger trampings about the Virginia woods were swarming like ants on an armadillo-free island in Venezuela. To many hunters and wildlife professionals, the crowded forest was an American success story. For those with broader concerns for such things as flowers and birds and future generations of forests, the deer resembled more the apocryphal fire bearing down on a sleeping village.

Leopold

One of the first to raise the alarm was the same man who had helped light the tinder. In 1933, a University of Wisconsin wildlife professor named Aldo Leopold authored *Game Management*, the first textbook dedicated to managing and restoring wildlife populations. Still a cornerstone in the field, *Game Management* pivots on the philosophy that "the way to manage game is to manage habitat." And the way to manage deer is to clear.

With his disciples storming the woods with their saws, the father of wildlife management began to suffer second thoughts. Those well-lit gaps with their flushes of green food had indeed stoked the herds' reproductive engines as planned. Though with the herds' predators now gone, nobody had given enough thought as to how to turn them off. All that energy streaming from the sun, much of which had once gone toward growing the forest's next generation, was now channeling into the bellies and biomass of the burgeoning deer herds.

Invited to Germany in 1935, Leopold toured the country's forests, meticulously managed for maximum timber and game. To Leopold, there was something eerie about the German superforests, something bleak in their factory productions. The images reminded him of a dark Devonian forest from the predawn of the dinosaurs. There was spruce in the canopy, ferns on the ground, and little else in between. He wrote, "One cannot travel many days in the German forests, either public or private, without being overwhelmed by the fact that artificialized game management and artificialized forestry tend to destroy each other."

Leopold returned to the States a changed man. He had seen in the sterility of the German woods a harbinger of the Wisconsin forest. From his chair as a Wisconsin conservation commissioner, Leopold became the forest's spokesman, campaigning for more deer hunters and mercy bullets, in lieu of stripped forests and slow death by starvation. Leopold even ventured an olive branch to the timber wolf and mountain lion, the same vermin whose exterminations he had once championed as a brash tenderfoot forester. He warned that Wisconsin was heading toward a bitter end, "with impoverished herds, depleted forests, and (I hope) a fund of painful experience."

Time and again Leopold's message was shouted down by mobs of hunters and their politicians, teeth bared in defense of the swelling herds. Leopold died in 1948, with the game farmers in power and the deer herds on moonward trajectories. But he had by then sown the seeds of resistance. He had started his students and their students thinking more deeply about the forests as something other than feedlots for white-tailed deer.

"Dandelions Are Nice"

In 1985, three botany students at the University of Wisconsin-Madison got together over home-baked chicken cacciatore and a fresh draft of Wisconsin's new forest-management plan. The three had just finished several seasons in the woods, surveying rare plants throughout northern Wisconsin. What the U.S. Forest Service was proposing for those woods struck the young botanists as lunacy. The feds were planning more clear-cuts, to harvest more timber, to grow more deer. Everything the botanists had seen suggested that more deer were the last thing the forests of Wisconsin needed.

In their forays they had so often come upon the beheaded stems of lady's slipper orchids and trillium lilies, growing ever more rare in forests as sterile as city parks. They were astonished, by way of contrast, by the flourishing forests they found on the Menominee Indian Reservation—where liberal hunting seasons had kept the deer herd trimmed to a fraction of those found in the national forests. The cedars of the reservation were like none they had ever seen. "Since I was a kid, I had always thought of cedar swamps as being these beautiful parklike things where the understory is really clear, where you could see a long ways, where you could spot orchids thirty or forty feet away," said the botanist William Alverson. "And I got into the Menominee and realized I couldn't see more than five feet. It was because of the cedar regenerating. I said, 'Holy crap, this is what a cedar swamp is supposed to look like!' "

Alverson and colleague Stephen Solheim enlisted Solheim's adviser, Donald Waller, to their cause. Waller was a botany professor, otherwise busy at that time "investigating uncontroversial topics like why plants have sex." Years earlier, as a graduate student at Princeton, Waller and his botany professor John Terborgh had conducted a study just outside the D.C. Beltway, in a park near Terborgh's boyhood home. They counted as many as two dozen species of native flowering plants in a single square meter of forest floor along the Potomac. The same abundant forest that had sensitized Terborgh's eye to the decay that followed had served Waller's as well. "I mean, here we were in the land of Leopold," said Waller. "Here we were beginning to see the dramatic effects of deer in the woods, and nobody in wildlife ecology, nobody in the Division of Natural Resources, was researching any of this."

After the foresters blithely waved away the botanists' recommendations, the botanists went public. Their resulting paper, "Forests Too Deer: Edge Effects in Northern Wisconsin," appearing in a 1988 issue of the new journal *Conservation Biology*, served as both a warning to their colleagues and a bald indictment of the deer managers. The three buttressed their own observations in the Wisconsin woods with evidence marshaled from across the scientific literature, citing studies from Nebraska to Pennsylvania to Long Island. Missing species, failing forests, outbreaks of Lyme disease—all came under one suspicious shadow of supersaturated densities of deer. Stewards of rare orchids had to cage their plants against the swarms. Commercial foresters in Pennsylvania had resorted to fencing off their stands as a pre-

requisite for growing new timber. And the U.S. Forest Service was prescrib-
ing more deer?

In the year following publication of "Forests Too Deer," the authors received upward of a thousand requests for reprints. Few of them came from
the deer managers. "I was told at the time that we were being alarmist,"
said Waller. "I was told the problems were local, temporary, that it's only a
matter of time before a harsh winter would knock back the deer population,
or that the hemlock wasn't regenerating because of climate change. Whatever."

Ten years later, Waller and Alverson were back in print with the same
warning in sterner tones, loaded with reams of more damning evidence of
a situation inexcusably worsening. Conducting their own surveys throughout the hemlock and cedar woods of northern Wisconsin, they found aging
trees and understories barren of seedlings, the forests slowly dying. They set
up fences to exclude the deer and watched those areas sprout jungles while
outside remained stubbled pasture. They cited a study listing ninety-eight
species of rare plants of concern being devoured by deer. Once they started
looking for the deer's impacts, they found them everywhere.

One of the more poignant images came from a rare stand of virgin forest
in the Allegheny Plateau of northwest Pennsylvania. Heart's Content National Forest Scenic Area was a 120-acre grove of hemlocks and white
pine—some of them four centuries old—a tiny island of giant trees surrounded by adolescent stands of cropped timber. Heart's Content was a state
preserve—no logging, no hunting, no tampering with nature; a supposed
port in the storm—yet the bastion had come to resemble more the battle-
field. In 1929, a prominent ecologist named H. J. Lutz had surveyed the
plants of Heart's Content. By the mid-'80s, half of the tree species recorded
by Lutz were declared missing. A decade later, another survey tallied upward
of 80 percent of the understory flowers missing as well. The forest floor had
been nearly swallowed in a sea of hay-scented ferns—one of the few plants
no white-tailed deer would touch. Heart's Content, in all its purported purity, had become a graveyard of towering tombstones waiting to topple.

It took an extreme bit of maneuvering for an edible plant to find refuge in the modern forests of Pennsylvania. Thomas Rooney, the young ecologist who had documented the understory decay of Heart's Content, had noticed in a nearby forest the ground cover increasingly sparse of one of its most previously common wildflowers, a species of deer candy called the Canada lily. Rooney soon discovered, however, he could still find rich pockets of the lilies, if he was willing to climb for them. The last of the lilies had taken refuge atop boulders, upward of fifty feet above the ground.

Rooney soon thereafter came to study under Waller at the University of Wisconsin, and the two set about charting a comprehensive reckoning of the forests' demise. They began on campus, in a storeroom of old filing cabinets housing the notes of the eminent botanist John Curtis. In preparing his 1959 classic, *The Vegetation of Wisconsin*, Curtis had carried out the Herculean task of systematically canvassing the entire state for its plant communities. For Waller and Rooney, Curtis's notes became their Rosetta stone.

Retracing Curtis's steps, Waller and Rooney, along with colleagues Shannon Wiegmann and David Rogers, chronicled fifty years of change in the Wisconsin flora. "Curtis provided us a much richer picture of what's going on across the landscape," said Waller. "And it was a disturbing picture."

Across the forests of northern Wisconsin, a fifth of the understory's native species had gone missing since the surveys of Curtis. But the percentages only began to describe the damage. Habitat generalists had prospered, and habitat specialists had declined. Or by Waller's analogy, "Dandelions are nice, and my daughter likes to pick them. But it's like junk food—you don't want to see it or eat it all the time." The impoverished deer forests of northern Wisconsin—in a trend that echoed across the American East—had by the latest measures become botanical strip malls, monotonous repetitions of deer-proof weeds, sprawling in the wake of the herds.

Haida Gwai

Waller and his colleagues could only speculate beyond their botany to the animals whose fates they suspected were intimately entwined with the failing forests. It was especially worrisome that the hardest-hit flowers—the trilliums and lilies and lady's slippers, with their conspicuous blooms and ice-cream appeal to whitetails—were also those most heavily reliant on insects to carry their pollen, and birds and mammals to disperse their seeds. It remained to be seen how much the animals would in turn miss their flowers.

As it happened, a French ecologist named Jean-Louis Martin was already on the case, albeit on the far side of the continent in a temperate rainforest off the coast of British Columbia. In 1993, Martin had first come to the forests of Haida Gwai—an archipelago of some 350 islands clustered fifty miles from the Canadian mainland—with some basic questions on birds and the theory of island biogeography. The theory is a sacred formula, proposed in 1963 by the eminent ecologists Robert H. MacArthur and Edward O. Wilson to explain why small and distant islands tend to hold fewer species than others. Leaving the larger relevance of island biogeography for a later discussion, the salient observation here involves Martin's arrival on Haida Gwai, where he immediately witnessed the sacred theory violated. Touring dozens of islands big and small, Martin found the most birds— oddly enough—on the smallest, most remote islands. It was soon plain to see why. The forests of the little islands were verdant jungles compared with the moss-carpeted parks of the bigger islands. The little islands were thick with bird habitat. They were also, most conspicuously, the only islands free of deer.

For ten thousand years, the forests of Haida Gwai had evolved free of browsing deer. Within the past one hundred years, mainlanders loosed some black-tailed deer on the islands and inadvertently began an untended experiment in the ecology of overabundance.

Freed from the wolves, bears, mountain lions, and hunting humans otherwise prowling the mainland, the blacktails found life especially plush on Haida Gwai. As each little Eden got a bit too crowded, with blacktail densities approaching fifty per square mile, a deer or two would swim to the next virgin paradise beckoning across the channel, and so forth, until they'd colonized all but a handful of Haida Gwai's islands.

Martin's team documented not only the demise of Haida Gwai's understory, but also the birds and insects that went with it. Islands browsed for more than fifty years had lost up to three of every four species of their songbirds. Insect species plummeted sixfold. Many of them were pollinators, raising the possibility that if the deer hadn't undone Haida Gwai's understory, sterility might have. The undoing of Haida Gwai caused the researchers to wonder whether deer might be to blame for the broader ongoing collapses of North America's migrant songbirds and pollinating insects.

Perhaps more than any numerical effect the missing wolves might have had on the deer was their psychological impact. In this land of no predators, the deer grazed as boldly as bulls in summer pasture. Two days after setting up camp, Martin's crew had their subjects eating out of their hands. "For me it was sort of a major lightbulb which came on," said Martin. "Suddenly what I realized working there [is] that carnivores are mainly not animals which eat prey but which change the behavior of prey." And such a change could trigger biological tremors to the base of the food chain.

"We can now say," Martin continued, "that limiting predators has costs—costs to forestry, and costs to wildlife. If you want to protect land without wolves, you risk losing birds, plants, et cetera. Wolves can thus be management tools. That may sound shocking, but the more one examines our work, the more we can say that this could be a solution, the only solution in some areas, to the problem caused by overabundance of deer."

Out of Wisconsin

The most striking feature of such revelations as Martin's and Waller's were their ubiquity. In the mid-1990s, more than forty biologists and wildlife historians gathered for a symposium in Washington, D.C., with deer stories of their own. The result was a book, The Science of Overabundance, which ran 402 pages long.

William McShea, the lead editor and organizer of Overabundance, had been charting his own slow-moving catastrophe in the Shenandoah Mountains of Virginia. It wasn't just oak seedlings taking a beating in McShea's plots.

Gray squirrels, flying squirrels, chipmunks, and a host of small mammals of the Blue Ridge surged back wherever McShea fenced the deer out. With the understories also went the birds. McShea could confidently predict the disappearance of certain songbirds by the amount of time the deer were allowed to overrun the birds' habitat.

Researchers in the Alleghenies of northwestern Pennsylvania found likewise. Where the deer surpassed twenty per square mile, the bird diversity dropped by almost a third. Songbirds like the indigo bunting, least flycatcher, eastern wood peewee, yellow-billed cuckoo, and cerulean warbler just disappeared.

In places, even the predators were no longer safe. Anticosti Island, off the coast of Quebec, had once been famously thick in black bears, fattening themselves off huge crops of gooseberries and currants. Within one hundred years of the white-tailed deer's introduction, there were some 150,000 whitetails on Anticosti, and no more berries. No more berries, no more bears. Remarked Anticosti's researcher, Steeve Côté, "To my knowledge, this is the first evidence of what appears to be the indirect extirpation of an abundant large carnivore by an introduced herbivore."

Time and again, the heaviest damages visited forests where the carnivores were gone, the hunters had been barred, and the only remaining predator was the automobile. It was a situation that happened to describe the charter of most national parks. In the Big Meadows wetland of Shenandoah National Park, a herd of deer counted at densities of 150 to 200 per square mile, and fattened on campers' Cheetos and potato chips, had the park biologists scrambling to rescue what was left of a rare plant community found nowhere else on Earth. Said Shenandoah wildlife biologist Rolf Gubler, "We've created a monster."

Cades Cove in Tennessee, population epicenter of both tourists and deer of the Great Smoky Mountains National Park, receives more than two million visitors each year, few of whom have any idea what they're missing. In the 1940s, when the park was established, Cades Cove was known for its huge displays of trillium and spring wildflowers. Modern botanists are hard-pressed to find trilliums there now. Forty-six species of wildflowers recorded in Cades Cove in 1970 were gone by 2004, all of them relished by white-tailed deer.

Fence and Fortress

Hidden deep in the Monongahela National Forest of West Virginia, in an otherwise undistinguished hollow (for discretion's purpose, heretofore called the Hollow), stands what might represent the future of rare orchid conservation in America. There, one of the state's last two patches of *Cypripedium reginae*, the showy lady's slipper orchid, grows in what amounts to a terrestrial shark cage, surrounded as they are by an eight-foot-tall deer fence.

The showy lady's slipper orchid is one of the biggest, most extraordinary members of what many aficionados consider the most alluring family of flowering plants. With striking white and pink blossoms atop stems reaching to the thighs, the showy lady's slipper is the queen of North American orchids. It is a rather widely but precariously distributed orchid, once found more commonly throughout the northeastern United States but lately rare and disappearing at worrisome rates in its southernmost locations. When orchid specialist Kathy Gregg from West Virginia Wesleyan College began visiting the lady's slippers of the Hollow, she discovered the population in the midst of a free fall. In 1987, Gregg counted 650 orchid stems in the little clearing. Later that year, deer had eaten 95 percent of them. The next year the orchids gamely sent up another flush of flowers. And again the deer mowed them down.

In 1989, the fence went up. But the orchids had been deeply wounded. Three straight years of intense cropping had sapped their underground reserves. Convalescing inside their cage, the lady's slippers took a decade to fully recover. "Without that cage, I don't think they'd be here," said Gregg. "The deer just mowed them, eating them to the ground. If it had gone on much longer, the population would have been gone."

With deer now the reigning predator of the woods, life for the showy lady's slipper has become a perilous existence. Gregg reveals that not all uncaged lady's slippers are getting chewed by deer. In fact a free-growing population remains farther south, in Tennessee. The botanist who studies them does so by rappelling down the walls of the cliff on which they grow.

Must We Shoot Deer?

In 1992, the biologist Jared Diamond penned a famous essay entitled "Must We Shoot Deer to Save Nature?" Diamond had spent a "magically beautiful, but painfully upsetting, day in Fontenelle Forest near Omaha." A thirteen-hundred-acre tract of Missouri River floodplain forest, Fontenelle is one of the few surviving forests along banks dominated by suburbs and farmlands. The intent to protect the forest from human harm—no hunting, no fishing, no flower-picking—had produced instead a macabre museum of the living dead. Diamond saw old oaks and hickories but no seedlings. Their understory nurseries had become "a haven for 'deer-proof' plants, such as poisonous snakeroot and stinging nettles." Birds and butterflies were disappearing. "The sight felt like visiting an apparently thriving country and suddenly realizing that it was inhabited mainly by old people, and that most of the infants and children had died," Diamond wrote.

Our fault, confessed Diamond. "We provide the equivalent of supermarket conditions for deer by breaking the landscape into the habitat mosaics that they prefer, planting crops on which they feed, and eliminating the big carnivores that used to keep down their numbers." Without those key predators patrolling the food web, he said, "what we're trying to preserve is no longer the pristine self-sustaining ecosystem that nature could manage unassisted, but an already collapsing ecosystem incapable of sustaining itself."

"It's easy for me," Diamond excused himself in closing. "I'm not the one who will have to explain to the public why they can't pick flowers in a reserve where deer are shot."

Diamond was wise to count his blessings. Within the decade the shooting had indeed started, in both directions. With some of the last bastions of biological diversity under plague of the hoofed locusts, conservationists started killing. The Audubon Society, better known for hunting birds with binoculars, went on record in favor of gunning overstocked deer; the Nature Conservancy, manager of the largest private system of nature reserves in the world, started opening some of its most heavily besieged sanctuaries to crews of archers and riflemen. Antihunters and animal rights activists fired back with protests, media campaigns, and op-ed grenades suggesting the hunts were merely cruel excuses to kill.

But some of the most vehement opposition came from hunters themselves, for whom there could never be too many deer. And nobody felt their wrath like Gary Alt. In the year 2000, after a long wildlife career spent crawling headfirst into bear dens, the biologist Alt enlisted for a new post on the Pennsylvania Game Commission to save the forest from its deer. The forests of Pennsylvania, browsed by the largest deer herd in the East—to the chagrin of foresters and conservation biologists, and to the delight of Pennsylvania deer hunters—had entered a state of reproductive arrest. (Among the few trees still regenerating was black cherry, a tree with cyanide in its tissues.) Alt's first task was to deliver the bad news to the public, among them a powerful culture of hunters beholden to their behemoth herds. Alt soon came to long again for the safety of a bear den.

Night after night, in a 225-town tour of Pennsylvania, Alt played to overflowing auditoriums, projecting on big screens the damning photos of fence lines, delivering the same dark sermon. "If you care about forest ecosystems, if you care about wildlife, if you care about hunting or just care about deer, this ought to drive you to your knees," Alt would say, pointing to the deer-beaten side of the fence. "You can't grow deer in this habitat on the left. For seventy years we have had more deer in that system than we should have. Healthy deer can double their numbers in two years. Devastated habitat may take decades to recover."

The obvious remedy to Alt and his fellow biologists was heresy to the Pennsylvania deer hunter. For suggesting more deer needed to be shot, Alt was sworn at, spat on, and threatened with death. He began reporting for work in a bulletproof vest. His crusade eventually cost him his marriage and his job. After twenty-eight years with the Pennsylvania game department, Alt resigned, moved to California, and took up photography and leading natural history tours. A happier Alt later said, "I would rather walk into a grizzly den unarmed than try to sell deer management in the state of Pennsylvania."

Coming Soon to a Neighborhood Near You

Huntley Meadows Park is an ancient meander of the Potomac River south of the District of Columbia, encompassing two square miles of oak woods and wetland meadows, with a population of white-tailed deer that not long ago numbered 250 (roughly twenty times the park's ecological carrying capacity). When Gary Roisum came on as park manager in 1978, there was already a browse line forming in the understory, and the ground cover was growing noticeably sparse. In the mid-1980s, Roisum noticed a patch of exotic grass growing near the visitor center. Microstegium vimineum, Japanese stilt grass, is an immigrant from Asia first spotted in Tennessee in 1919, supposedly arriving in a crate of porcelain, which it was used to pack. Roisum would come to wish he'd yanked that harmless little patch when he had the power to do so.

By the 1960s, Japanese stilt grass had appeared in all the Atlantic states from New Jersey south. As the conservation community caught wind of it, Microstegium emerged as no run-of-the-mill garden weed. A single plant could produce a thousand seeds; it could survive ten weeks underwater and grow to maturity in the weakest light of the forest. Microstegium, it turned out, could alter the soil chemistry to the demise of its competitors. And whatever Microstegium couldn't overcome on its own, with whitetails they could. With their hooves the whitetails prepped the stilt grass's soil, with their incisors they cleared the competition. Roisum watched as each season the green wave of stilt grass rolled farther across his sanctuary, spreading in a front like a slow-moving fire. He pleaded for defense funds that never came. Together, deer and stilt grass blazed through Huntley Meadows, immolating the spring beauties and violets and pink lady's slippers along the way. The deer browsed the high-bush blueberry to stubs. No seedlings of hickory, ash, oak, or sassafras reached beyond the molar zone of a white-tailed deer. Roisum's park was under attack.

And soon, so was Roisum. One day at work in the late 1990s, Roisum fell ill in a hurry. His temperature leaped to 103.5 degrees Fahrenheit. He tried to drive home, but reached only so far as the emergency room. For the next three days doctors marinated Roisum's immobile body with antibiotic cocktails. Roisum had been diagnosed with Lyme disease, an

incapacitating bacterial infection transmitted by a tick—a tick once re-
ferred to as the deer tick.

Since its emergence in 1975, Lyme disease—named for the town hosting
one of the first outbreaks—has spread from its Connecticut bull's-eye across
the forested East. It sprang up in the Great Lakes region, in the forests of the
West Coast. The U.S. cases have reached fifteen thousand to twenty thousand
per year—an underestimate by a factor of ten, according to some. Lyme has
vaulted to the rank of fastest-growing infectious disease in the United
States.

Traced to an infection by *Borrelia burgdorferi*, a spirochete bacterium related
to the syphilis germ, Lyme, if left untreated sometimes leaves its victims
double-visioned and seizing, with their skin burning as if on fire. Doctors
mistake Lyme for chronic fatigue syndrome, rheumatoid arthritis, multiple
sclerosis. Lyme, the great masquerader, commonly goes misdiagnosed and
untreated. In its later stages the disease produces meningitis and facial paral-
ysis, hallucinations and panic attacks. Long-term victims sometimes wind
up in wheelchairs, some in the grips of crippling pain, pondering suicide.
And it seems that wherever Lyme erupts, deer are there in numbers.

Now thirty years beyond the first outbreak, there comes a new twist to
Lyme's baffling ascent to the ranks of public enemy. A once simplistic equat-
ing of too many deer with outbreaks of the disease has since grown complicated. It is indeed true that a single deer may harbor thousands of adult
black-legged ticks whose species indeed harbors the Lyme germ. But other
accomplices have emerged. For it is not the adult ticks, rather their tinier
nymphal offspring that tend to acquire and pass the *Borrelia* bacterium to the
human bloodstream. And those dangerous little nymphs commonly get
their doses of *Borrelia* by sipping blood from chipmunks and white-footed
mice, two otherwise handsome and ubiquitous denizens of the forest understory. It now appears that one can better predict a hot zone of Lyme by
censusing its mice.

Much of the above has been informed over the last twenty years by the work of Richard Ostfeld and his colleagues at the Institute of Ecosystem Studies in Millbrook, New York. And though convinced of the mouse's giant role in harboring the epidemic, Ostfeld himself is not ready to acquit the deer. Ostfeld's hypothesis holds that in precolonial times, both deer and ticks were far fewer than today. Sometime in the mid- to late-twentieth century the rebounding deer herd crossed a population threshold, triggering a historically tiny population of ticks to turn abnormally huge. As much as mice may have harbored Lyme's bacterial reservoir, it was the defanging of America and the ensuing eruption of deer that sent the legions of transmitting ticks swarming the mice's way, and on up the pant legs of the American populace.

Head-on

When he recovered, Gary Roisum went back to Huntley Meadows to again face the vandals holding court in his castle. His predicament was nothing if not common. Almost every park manager in Fairfax County was by then reporting to work with a cud-chewing boss in charge. The science of deer management had not yet devised a workable system of chemical contraception. Translocations had turned into costly disasters, with the captured deer dying in droves, never mind the awkward question of where on earth any surplus deer were supposed to go. To the managers' minds, there was only one practical option left, the mention of which invariably sent the local politicians ducking for cover. "Every time we'd bring it up to the board of supervisors," said Roisum, "everybody was afraid we were going to be turning a bunch of guys loose with a shotgun and a six-pack to go after deer."

On an October morning in 1997, Roisum's lethal rationale gained a horrific measure of leverage. That morning, Sheryl Czepluch, a forty-nine-year-old Fairfax school librarian, was driving to work along a suburban two-lane road when up ahead a deer jumped the guardrail. The deer bounded onto the asphalt and into the path of an oncoming Mercedes-Benz. The Mercedes launched the deer through the windshield of Czepluch's Volvo. Czepluch's head and chest bore the impact of a hundred-pound sledgehammer. Her collision was one of a million and a half nationwide, her death by deer added to the conservatively estimated 150 to 200 human casualties every year. Soon thereafter, Fairfax began shooting its deer.

The county hired veteran wildlife biologist Earl Hodnett out of retirement, and he launched one of the country's most sophisticated guerrilla campaigns against the suburban deer. In the blackness of night, in trucks painted olive drab, Hodnett's crews of police sharpshooters began rolling into the deer-plagued parks of Fairfax County. Helicopters hovered overhead, searchlights shooing romantic couples and loafing drunks clutching vodka bottles. From the back of the trucks, the sharpshooters stood armed with high-powered rifles fitted with silencers, scanning the woods with infrared night vision scopes, freezing their targets in the beams, and delivering each a single shot to the head.

Since the shooting began, the deer numbers have started to drop, from the obscene to the merely stupendous. Aiming to eventually stabilize the deer population at fifteen to twenty per square mile, the job of Hodnett has become that of the Dutch boy plugging the dike.

Beyond the parks, Hodnett's jurisdiction covers the backyard battleground, where punch-drunk gardeners have resorted to rigging electric scarecrows, spreading lion dung, peeing in their patches, and ultimately dropping their trowels in defeat. It is Hodnett who hears about it when the king of a gated castle in the posh woods of McLean loses ten thousand dollars in custom landscaping to the local gang of deer. It is Hodnett who fields the complaints about audacious deer climbing three-story decks to reach the flower boxes.

Hodnett has also become Fairfax's de facto epidemiologist. "People call me complaining one of their kids has Lyme disease. And their dad has it. And the neighbor across the street and another neighbor down the street has it," said Hodnett. "They'll rattle off a dozen people on their street that either had it this year or over the last couple years. They complain to me as if I have the ultimate power to take care of the deer problem."

Hodnett is Fairfax County's sole wildlife biologist—one man versus fifty thousand deer. "I'd say seventy-five percent of my job is deer," he said, noting that the deer have lately added reinforcements. "The remaining quarter is Canada geese and beavers and creatures in the attic making noise."

It is perhaps more than coincidence that this new suite of pests now flooding Hodnett's to-do list were once—like the white-tailed deer—more commonly considered prey.

SEVEN Little Monsters' Ball

IN THE LATE 1970S, ornithologists and amateur bird-watchers of the eastern United States began to grow uneasy about a certain hush sweeping over their favorite woods. It seemed the summer chorus of songbirds was falling fainter by the year.

They could not pass it off as a dip in some recurring cycle, or an inflated memory of the good old days. Shortly after World War II, a volunteer corps of bird-watchers from the Audubon Society started formalizing their weekend outings with an annual census of breeding birds. The society faithfully compiled and published the surveys for the next forty years. What the birders felt in their guts, their census now confirmed as fact. A certain suite of migratory birds was in sustained and serious decline.

One of the longest-running censuses covered Rock Creek Park, a nine-mile corridor of forest snaking through the heart of the District of Columbia. Walled in by the embassies and apartment towers of urban Washington, Rock Creek Park appeared as one of the Mid-Atlantic's mightiest magnets for avian species funneling through the eastern megalopolis. In its postwar heyday, the park and a few of its urban green outliers had become the birding grounds for legends of the genre, Roger Tory Peterson and Rachel Carson among them. It was granted as fact among the binocular-clad cult that the highlight of springtime in Washington was not the blossoming of the cherry trees but the return of the warblers.

Now the return of the warblers could no longer be assumed. By the 1970s, the Kentucky warbler, northern parula warbler, black-and-white warbler, and hooded warbler were dropping off the charts. Fellow migrants such as the Eastern wood-pewee, Acadian flycatcher, scarlet tanager, ovenbird, red-eyed vireo, and wood thrush were down by half. By 1986, the sum of all breeding pairs in Rock Creek had plummeted by two thirds.

Scientists weighed in with their best guesses. There was always that widening black hole of tropical deforestation. With every autumn, the bulk of the migrants would funnel south into Central America, to an area of land far smaller, into forests felled more rapidly, than the breeding grounds to which the birds returned every spring. Yet the troubles in the tropics could not explain the birds disappearing from certain northern forests and not others. In the giant unbroken tracts of the West, there seemed to be no problem. It was mainly east of the Mississippi, where the forests were being chopped and diced, that the birds were most rapidly disappearing. As much as tropical deforestation loomed as an undying threat, something equally worrisome and far murkier was descending on the northern breeding grounds.

The North American concerns came to focus on fragmentation, the process and end-product by which vast swaths of forest were being cut into smaller isolated woodlots. The forests had become habitat islands. And for a budding generation of conservation biologists, islands had become the top-most topic of fascination and fear.

Their fixation was fueled in large part by Robert H. MacArthur and Edward O. Wilson's pioneering theory of island biogeography, offering an explanation for the life and death of species, as exemplified through the faunas and floras of oceanic islands. The theory states, here in extreme shorthand, the deceptively simple observation that the smaller and more isolated the island, the fewer the number of species it is ultimately apt to harbor. The diversity harbored by the island eventually arrives at a balance point between colliding forces. Those main forces are extinction and immigration—the chance of extinction growing greater, the chance of immigration growing smaller, as the islands shrink, and the mainland disappears over the far horizon.

That, for the biologist, was the fascinating part. The fear, for the conservationist, came from speculating how well the theory might hold for the mainland, how well it might predict the demise of many species living on what, in essence, were mainland islands. The evaporating forests of the eastern United States were one in a list of ecosystems that had been set adrift between widening gulfs of asphalt, concrete, brick, and steel. It seemed reasonable to expect similar problems facing inhabitants of such lonely little woodlots as with those on specks of land far upon the sea. As living space decreased, the islanders' numbers would ratchet down in step. At some point the little group's survival would become an issue. All other things being equal, smaller populations would ultimately be less resilient to the various bumps and potholes of life. Drought and plague, fire and flood—jolts enough to merely wobble a robust population—could topple those of tinier numbers. And such a little clan of castaways, so far removed from the motherland, was less likely to be rescued by replacements happening across the formidable gulf.

It was such analogous speculation that had the science-minded birdwatchers of places like Rock Creek wondering where the bottom might lie for their embattled island of birds. If there was nothing to be done about it—given the prospects of reverting the nation's capital to temperate deciduous forest—there was at least the duty to determine the causes of death. The interrogations turned back again to fragmentation, and the ecology of life on the ever ubiquitous edge. Tests would confirm that it wasn't the physical edge per se that was draining the little migrants. The birds had no measurable hesitance to nesting right up to the forest's brink. The edge itself was innocent enough. But the same could not be assumed of that suspicious cast of predatory characters that had lately come to lurk there.

Empty Nests

In the early 1980s, finding his songbirds in free fall, a Princeton student under the tutelage of John Terborgh took a stab at fingering the cause. From the beginning, David Wilcove's leading suspect was predation.

Wilcove had recently become aware of the experiment on Panama's Barro Colorado Island, where the bird species had been blinking out since the island's formation sixty years before. On the island, Bette Loiselle and William Hoppes, two graduate students from the University of Illinois, had

set out lines of wicker nests filled with commercial quail eggs, and returned
to find most of them raided. It struck Wilcove—as it had his adviser,
Terborgh—that the dropping bird counts of Barro Colorado uncannily reflected those from his homeland forests in the eastern United States. There
was an added hint of coincidence in that the chief suspect for the egg bur-
glaries on Barro Colorado was the coati, whose North American cousin, the
raccoon, was at that time multiplying to unheard of numbers.

And the raccoons weren't alone. The backyard menagerie of modern
America had come to include a glut of blue jays, crows, squirrels, and opossums, scuttling among pet dogs and house cats by the hundreds of millions.
With wolves and cougars and eagles no longer patrolling what was left of
the great American forest, the lower-ranking predators had been granted
free reign. And all came well fed by an endless gravy train of pet food and
bird seed, cornfields and garbage cans. (The raccoons lording over Long
Island were reported to have doubled in body size.) To Wilcove, the subsidized, fearless fauna of suburbia appeared as something more than a motley
bunch of beggars. Wilcove saw in each a potential predator of the missing
songbirds.

For his test sites, Wilcove sought out extremes on the ecological spectrum. A half hour's drive north of Washington, D.C., he singled out a set of
isolated forests from the suburbs and farmland of Maryland, some as small
as ten acres. These were his island analogues. For his mainland counterpart,
Wilcove ventured four hundred miles south, to the half-million-acre Great
Smoky Mountains National Park, refuge of the biggest tract of virgin forest
in the East.

In the fashion of Loiselle and Hoppes, Wilcove laid out lines of wicker
nests through his woods, from edge to forest interior. He placed some at the
base of trees and shrubs, affixed others on the lower branches, simulating
the nesting habits of his species of concern. And like Loiselle and Hoppes,
he returned to find scenes of carnage.

Within a week, in the patchwork forests of Maryland, something had
plundered most of Wilcove's makeshift nests. And the closer to the civilization, and the nearer to the ground, the more complete the plunder. Seventy
percent of Wilcove's suburban nests lost eggs. In the dirt around those nests

he read the tracks of dogs, cats, raccoons, opossums, skunks, and blue jays. Just to make sure, Wilcove sat and waited in a blind. Within half an hour, a blue jay flew in and speared an unguarded egg.

Meanwhile, down deep in the spooky wilds of the Smokies, where the bears and the bobcats still ruled the woods, Wilcove's nests rested safely. All but two in a hundred remained unscathed at week's end.

Damning as they might have seemed, such numbers could only be trusted so far. Wilcove admittedly wasn't as crafty at hiding nests as his subjects were. One could imagine a particularly industrious Maryland raccoon clueing in and following his lines like stashes of Easter eggs painted in neon. Still, there was no waving away the acute contrasts between the sheltered wilds of the Smokies and the tamed and treacherous outskirts of Maryland. Life in the little woods, for certain little birds with low-lying nests, had become ten to fifty times more dangerous. That, concluded Wilcove, was a big reason why such pristine façades as Rock Creek Park had become so frighteningly quiet.

Another factor was a fellow songbird called the brown-headed cowbird. Once primarily a grassland bird, the cowbird had in earlier times practiced a living snagging insects flushed by the hooves of the bison herds. To procreate while otherwise keeping up with the moving feast, the cowbird had taken to laying its eggs in the nests of other birds, enough of which never caught on to the ruse, thereby raising cowbirds at the expense of their own nestlings. As the bison were replaced by cows, and the grasslands migrated eastward in the form of grain fields and pastures, the cowbirds followed along. They advanced on the front of the retreating forests, into the territories of songbirds unschooled in the parasitic ways of cowbirds. Woodland warblers rather suddenly found themselves scrambling to feed the begging maws of cowbird nestlings twice their own size.

And then there were the tropics at the other end of the line. The forests of Latin America, to which many of the migrants returned every fall and winter, were falling at spectacular rates. Guatemala, by way of example, was on pace to be all but denuded by the year 2025. Radar images tracking the clouds of migrants as they made their way north from the Yucatán Peninsula to the Gulf shores of the United States showed their numbers shrinking by half in one twenty-year period.

Wilcove's predators were the final straw. Given the squeeze of shrinking forests and the infestations of cowbirds on populations already hovering so dangerously close to the ground, the unprecedented swarming of the bottom-rung predators had become the unbearable weight. "We have rather profoundly changed the nature of predation as a natural phenomenon over much of this country," said Wilcove. "We have greatly increased the number of mid-sized predators, greatly decreased the number of large predators, and there's a host of ramifications that stem from those changes."

Keystone Coyote

It soon became apparent that this plague of little predators had long legs. By the mid-1980s, at about the time David Wilcove was publishing his ominous observations from the Eastern forests, a fellow conservation biologist on the other side of the continent—in an arid Southern California landscape poles asunder from the leafy green woods of the East—was inadvertently heading toward a similar discovery.

Michael Soulé had grown up in the coastal hill country of San Diego and had become a naturalist poking about the dry mesas and sharp canyons scruffily cloaked in sage and oak scrub—"those wonderful strips of native vegetation running through the city," as he called them. After stints as a biologist in Africa, a tenured professor at the University of California at San Diego, and a meditative life at a Los Angeles Zen center, Soulé returned to the San Diego chaparral and his beloved "home in the boonies," just in time to find the real estate developers planning a new city that would all but swallow his sanctuary.

"I realized it was unstoppable," Soulé said, "but I thought maybe they could do it in a way that could protect some of the conservation values and wildlife in the area. So I approached the developers and said, 'Look, there's a way to do this that probably won't cost you anything but will allow some of the wildlife to exist, just by making sure some of the connections in the canyons are maintained.'" He told them about island biogeography, about the inherent ecological dangers of isolation. The developers informed Soulé that island biogeography did not apply to southern California.

"They blew me off," said Soulé. "And I got mad. I said, screw you. I'll start a research project."

Soulé mobilized a team of students and faculty from the University of California at San Diego. They set out to document the biological fallout as the chaparral was carved to little bits. Soulé chose for his barometer a special group of birds.

The canyons harbored a suite of birds tightly tethered to the chaparral. The group was a varied lot, including the California thrasher, the California quail, two species of wren, a colorful sparrow called the spotted towhee, the greater roadrunner, and two active little insect gleaners, the wrentit and black-tailed gnatcatcher. The chaparral birds rarely flew far, and seldom foraged or bred elsewhere but in the native shrubbery. Wings notwithstanding, they were essentially stranded in their little pockets of scrub. As the steep-walled canyons lost ground to the earthmovers and home builders, the obligate birds of the chaparral were squeezed ever tighter amid their shrinking archipelago of habitat. It had apparently become a question of island biogeography as to how long the birds would survive in their little lots—how long before a brushfire incinerated their little sanctuaries, how long before inbreeding sent the population into a death spiral of genetic decay.

One of his first days of fieldwork, Soulé and team came to the fragment on Point Loma. It was a tiny patch, no more than two or three acres. Soulé had often explored there as a kid. He and his coworkers sat on the edge, listening and watching for the chaparral birds they were sure must be there. There was nothing. "And it just hit us," Soulé recalled. "Wow. It's real. We didn't know we'd see such profound effects."

As Soulé and company got down to business, they took stock of thirty-seven chaparral fragments scattered throughout the San Diego suburbs, the smallest of them amounting to less than a half acre, the largest about a hundred. For some, it had been only two years since the houses and fences and highways had sealed them off; for others, it was going on a century. Soulé fully expected to find the birds closely tracking the tenets of island biogeography, their numbers declining over time, most drastically as the size of the islands got smaller and the spaces between them larger.

At the last minute, Soulé decided to add one more variable to his analysis.

Anticipating his critics, he conceded to measure the impact of predation. "If I didn't look at predation," said Soulé, "the people would say, 'Well, how do you know foxes and cats and coyotes aren't killing off the songbirds?' So we had to look at least at the presences and absences of predators."

After two years of surveys, the results were tallied. Predictably enough, the patches were dropping species of chaparral birds over time, in line with the diminishing area of chaparral. But as Soulé ran down the analyses, an unexpected third factor leaped out of the data, striking him squarely between the eyes. The diversity of chaparral birds was higher in those patches inhabited by coyotes. And Soulé instantly understood why.

He recalled those unforgettable moments at his little house on the chaparral, when the cat door would suddenly explode. "Bang-bang-bang, as the cat came flying through, racing like he was being chased by the devil," said Soulé. "And he was. He was being chased by a coyote. My, we lost a lot of cats." The wisest cat Soulé recalls ever owning had spent most of its time on the roof.

The significance, as Soulé and team would later formally summarize for print, "suggests to us that coyotes are helping to control the smaller predators (including cats) in the canyons, possibly contributing to the maintenance of the native, chaparral avifauna."

The phenomenon behind Soulé's conclusion was the same phenomenon that John Terborgh had postulated for the demise of the birds by coatis on Barro Colorado Island, the same that his student David Wilcove had uncovered in the urbanized Maryland forests. By all appearances, in places where the dominant predators disappeared, a guild of smaller, mid-size predators (or mesopredators, as Soulé came to call them) took charge and rioted, multiplying by as much as tenfold. Ecologically speaking, when the top dog was away, the underdogs—and undercats—would play. "The phenomenon appears to be quite general," wrote Soulé and colleagues. "We refer to it as 'mesopredator release.'"

The mesopredators had indeed been released, as a quick scan of the continent revealed. In the prairie pothole region of the Dakotas—famed as the duck factory of North America—red foxes had by the 1980s taken to cleaning out duck nests. Throughout the shrinking forests of Illinois, an entire suite of ground-nesting songbird species was failing at dramatic rates, in suspicious concordance with a threefold surge in the raccoon population. Up and down the eastern seaboard, on the beaches and in the maritime forests of the Atlantic, shorebirds were being chased off by roaming dogs and house cats and urban gulls; colonies of thousands of terns and skimmers and herons and egrets were deserting en masse, put to flight by marauding raccoons and foxes.

Across the Atlantic emerged more outbreaks of second-string predators, none more frightful than the plague of baboons sweeping the poacher-ravaged reaches of sub-Saharan Africa. From the Ivory Coast to Kenya, in the expanding vacuum of missing lions and leopards, monstrous gangs of baboons had begun terrorizing the countryside. Free to wander when and where they pleased, the emboldened apes were to become Africa's chief crop raiders and ubiquitous thugs, mugging women and children for their food, breaking and entering houses, and slaughtering livestock and wildlife in crushing quantities. In hard-hit pockets of Uganda, kids were staying home from school to help guard the family's fields and flocks. The marauding apes, indulging their growing appetite for meat, began gang-tackling wild antelope, tearing them limb from limb. In the aftermath of the rioting baboons, scientists would find fellow primate societies annihilated, whole forests of bird nests plundered. Baboons were evicting hyenas from their kills. Having dethroned the lion, Africa had crowned a tyrannical new king of beasts.

Soulé's mesopredator release hypothesis suggested a likely explanation to many wildlife declines and disappearances sweeping across the land. For little populations already squeezed to the edge, the modern pestilence of subordinate predators was amounting to the unbearably heavy load.

But for Soulé, mesopredator release presented an equally intriguing flip side. Could mesopredator *restraint* actually rescue such species on the brink? Could it be, for example, that such a predator as the coyote, the most vilified creature in the history of America, was standing between chaparral birds and their extinction by mesopredators?

It wasn't long before tests of Soulé's suspicions began to gain scientific

credence from sites far beyond the California canyons. In the prairie potholes of the Dakotas, where the red foxes had been emptying duck nests
with tailspinning rapidity, an unforeseen return of the coyote was turning
things around. After decades of fanatical warfare against the coyote, in
which half a million of the song dogs were being shot, gassed, poisoned,
trapped, and strangled year after year, the 1970s came with rumblings of a
cease-fire. The shifting cultural climate brought lower fur prices and fewer
trappers; societal and scientific outcry led to President Richard M. Nixon's
1972 ban on public-land poisonings. In South Dakota, aerial gunners were
temporarily grounded.

The coyote, given its momentary stay of execution, wasted no time looking back. In the Dakotas their numbers surged. And with the return of the
coyote, the reign of the red fox ended. The coyote, less a hunter of eggs and
more a hunter of mice and rabbits—not to mention trespassing foxes—
actually made life safer for the ducks. Where the coyotes had come to rule,
the average nest success of the Dakota ducks jumped by fifteen percentage
points, enough to lift the nation's duck factory out of the red. The authors
of one key study even went so far as to prescribe coyotes as a means of duck
production—an eye-raising suggestion from a profession in which many
had invested careers in the coyote's eradication.

Meanwhile, in an expanse of short-grass prairie in western Texas, a wildlife
biologist named Scott Henke was himself questioning the conventional wisdom of killing off coyotes. Aerial gunners had swooped down and removed
half the coyotes over thirty-nine square miles of Henke's study site. Within
nine months of the shooting, the prairie was jumping with kangaroo rats, to
the demise of its eleven fellow rodent species. Black-tailed jackrabbits, on the
other hand, tripled in number. And to no surprise, the skunks, foxes, badgers,
and bobcats of the mesopredator clan also gained at the coyote's expense.
Henke's postmortem inventory left one to wonder if the coyote gunners
hadn't shot themselves in the foot. How many pounds of the rancher's cattle
forage had the booming jackrabbits nibbled away? How many eggs of the
quail hunters' quarry had the surging mesopredators gobbled in the coyote's
absence? These were the fuller range of questions and consequences Henke
implored others to ponder before shooting: "Biologists need to remember

that indirect effects are the rule rather than the exception in most ecosystems," he wrote.

Nearly a decade after his serendipitous discovery in the San Diego canyons, Soulé returned to his chaparral birds for a reexamination of the phenomenon he had made famous. This time, with a graduate student named Kevin Crooks, Soulé set his antennae more pointedly at the carnivore community. Crooks set out track stations and trip-set cameras, gathering footprints and incriminating photos. He and his assistants gathered scat as well, sifting through postdigested items on the carnivores' daily menu. And with an extra dose of suspicion, they ordered round-the-clock surveillance on the canyons' cats.

Crooks went around to the canyon's bordering homes, knocking on doors. "We're doing research down in the canyon. Do you have an outdoor cat?" he would say.

More often than not, the answer was yes.

"Do you let it run loose?"

Lots of nods.

"Would you mind if I put a radio collar on it?"

Domestic Violence

There was by this time ample reason for suspecting the household carnivore. Since its domestication four thousand years earlier by the Egyptians, Felis catus had become the world's secondmost fearsome predator (behind only the rat, which, to be taxonomically fair, is a generic name ascribed to at least three species). The domestic cat, from the backyard venturer to the abandoned pet gone feral, had over the past five hundred years spanned the continents and major islands of the world. Turned loose upon the global fauna of native rodents, rabbits, birds, snakes, and lizards, it had chalked up a conservative estimate of thirty-three extinctions and uncounted decimations.

When curious scientists first started taking serious note of the predatory habits of the presumably harmless house cat, the enormity of the massacres stunned them. In the British village of Bedfordshire, biologists Peter Churcher and John Lawton enlisted an entire community of cat owners as field assistants in what would become a landmark study of domestic predation. They went door-to-door with their feline census, tallying seventy-eight cats variably attached to 173 houses in the community. Churcher and Lawton instructed their amateur corps of wildlife scientists to bag the contents of any prey their cats brought home. By years' end, the villagers of Bedfordshire had proudly presented the carcasses and body parts of 1,100 little animals. Mammals were especially popular, the bags filling with mice, voles, and shrews, here the occasional rabbit, there a weasel. One cat specialized in slaying bats. An impressive sampling of birds rounded out the cats' take. House sparrows, thrushes, robins, and blackbirds were all brought home for the master of the house. Some of the more industrious cats were bagging a hundred bodies a year, and it was not hard to imagine at least that many more victims never quite made it home.

When Churcher and Lawton started extrapolating Bedfordshire's results across Britain, the quaint little survey grew morbidly huge. The cats' annual countrywide take came to seventy million mammals and birds. The ecological gravity of those numbers was lost on many of the killers' owners. When the scientists published their results, other cat owners proudly wrote in—some apparently with *Guinness Book* ambitions—bragging about yearly scores of up to four hundred animals.

Bedfordshire became a microcosm not only of Britain but of the world. Fed and sheltered by society, unhindered by disease or starvation, their killing dampened neither by satiation nor scarcity of prey—immune to the hardships faced by its competitors attempting an honest living in the wild—the cat as recreational hunter had become a formidable blight on both predator and prey. In the United States alone, there lived anywhere from sixty million to one hundred million cats, most of them household pets, with uncounted millions lurking in semiwild populations. In rural Wisconsin, free-ranging cats had exceeded one hundred per square mile, outnumbering by several times all the native foxes, raccoons, and skunks combined. There were cats in America that could put the Bedfordshire hunters to shame, some credited with one thousand kills in a year. When all were accounted for, U.S. house cats were each year dispatching upward of a billion mice, voles, and baby rabbits, plus hundreds of millions of birds.

And not all of those populating the hit lists were garden-variety creatures. On the dunes and sands of coastal Florida, the cats were plundering imperiled populations of beach mice and piping plovers. On Key Largo, a colony of five hundred feral cats had taken up residence within stalking range of an endangered species of wood rat.

On naïve island faunas around the world, the cats' consequences magnified. The house cat ambushed its way through rare rodents in the Caribbean and Baja and rare lizards of New Zealand. Before dispatching the last of the Guadalupe storm petrel, a small seabird nesting in burrows, cats were killing nearly half a million petrels per year. Before conservationists stepped in and killed the cats on Natividad Island off the coast of Mexico, the cats had been cropping a nesting colony of black-vented shearwaters by a thousand birds a month.

Now, the cats were toying with Michael Soulé's beloved islands of chaparral. In an average-size fragment of undeveloped canyon, sport-hunting house cats were retrieving close to two thousand mice, birds, and lizards per year, not to mention how many more they ate on the spot or left to the ants. With that kind of take, it no longer seemed so mysterious that over the past century, the chaparral fragments had suffered at least seventy-five extinctions of their obligate birds. Accounting only for the house cat's carnage, it was easy to figure why the birds of the chaparral were winking out like a cheap string of Christmas lights.

But here, at least in a few of the bigger tracts of canyons, something was standing in their way. What in Soulé's 1988 study had been a suggestive clue to the beneficent power of the coyote, in his new study soon hardened with organic data. One in every five coyote scats contained cat remains. One in every four feline radio collars ended up leading a researcher to a remnant of somebody's pet, typically half-buried beneath a shrub. The mere suggestion of coyotes on the prowl had nearly half the canyons' cat owners keeping their loved ones inside. And that, to many little native residents of the chaparral, amounted to salvation. Where the coyotes roamed, the cats ran scared, and the chaparral birds sang.

It was a conclusion with awkward implications for so many. The chaparral fauna was being drained into the maw of society's most popular predator, only to be spared by the graces of its most persecuted. Canis latrans, the coyote—scourge of delinquent cats, guardian of imperiled birds, benevolent alpha predator of the San Diego chaparral.

It was solid science with quirky mainstream appeal, appearing in the science journal Nature as well as the New York Times. And yet it was a story about to be eclipsed by orders of magnitude, by another of uncanny similarities and striking contrast. The new story involved the coyote's big brother, the wolf, bringing protection to a place as wild as the San Diego suburbs were domestic—a place called Yellowstone.

EIGHT Valley of Fear

IN ITS MEANDERING PATH across the northern tier of Yellowstone National Park, the Lamar River emerges from a mountain canyon upon a lofty valley of grass and sage. There it flows through what many eyes perceive as the premier wildlife panorama of the Lower 48. The Lamar Valley is the place to find the grizzly pawing through the summer grasses, to hear the bison bulls bashing heads in the September rut, to scan the sere valley plains and hillsides flashing yellow with the rumps of grazing elk. The coyote pouncing for voles in the grass, the pronghorn antelope galloping over the sage flats—the reasons the asphalt two-lane that snakes across the Lamar terrace so often crawls with sightseers—had little to do with Robert Beschta's astonishment on his first visit in the spring of 1996. A hydrologist from Oregon State University at Corvallis, Beschta had walked straight from the tour road to the banks of the Lamar, eyes fixed on the river itself. Or rather, from Beschta's frame of reference, what was left of it.

"I was dumbfounded how bad it was," said Beschta. "Just dumbfounded." The banks of the Lamar were barren, steep, and saw-edged. Soils that had been building for millennia had in recent decades been sluicing seaward with every rush of spring snowmelt and summer cloudburst. Few trees, no underbrush, no canopy, and no shade meant lost habitat for birds. It meant no more beavers that had once built ponds there, nor the flush of life that typically followed. "The stream," said Beschta, "was falling apart."

In better times and places, such streamside bands of life—the ecological community more formally known as riparian—otherwise constituted the bull's-eye of biodiversity in the arid West. Composing a mere 1 percent of the landmass, riparian ecosystems harbored 80 percent of the West's faunal diversity. That at least was the healthy riparian's reputation. The river Beschta was watching in Yellowstone was something decidedly different. Something had happened to the Lamar over the last century that hadn't happened there for many years before.

Beschta and many others believed that something was elk. The Northern Range of Yellowstone had for decades harbored one of the densest, most pampered populations of unfenced elk on the planet. *Cervus elaphus*, is an open-ground grazer and giant member of the deer family; some of the more massive bulls top seven hundred pounds. For nearly a century in Yellowstone they'd been protected from hunters and stripped of their native predators by federally backed trappers. The elk had responded by filling the Northern Range with a winter herd of twenty thousand, one of the densest populations known, whose ecological footprint had crushed the river bottoms and upland aspen groves, where the shoots and saplings of the forest's next generation were perpetually pruned to the ankles.

In his travels Beschta had seen the same sort of dissolution throughout the overstocked range of the American West, where more than a century's excess of cattle and sheep had pounded so many of the arid land's foremost oases to dust. It had become so wearisomely familiar: mile after mile of rubbled streamsides, willow thickets rendered to nubs, denuded banks calving like miniature glaciers, the birds and beavers of yore all but gone. But it was especially unsettling to find the same sickness here in the heart of the nation's flagship bastion of nature.

What Beschta could not see, even then as he stood vowing to return someday and unveil the unraveling of the Lamar, was a new order about to descend on Yellowstone's ecology.

Coming Home

Seventy years before, near the spot where Beschta stood, two young wolves had simultaneously stepped, side by side, into steel-jawed traps. They were the last of at least 136 Yellowstone wolves killed during the nation's eradication campaign. With their elimination in 1926, Yellowstone had become wolf-less for perhaps the first time in the twelve thousand years since *Canis lupus* trotted in behind the retreating front of the Pleistocene glaciers.

Soon after ridding themselves of the wolf, the U.S. Park Service discovered another pest on their hands. Elk, the park's antlered showpieces, amassed like locusts, chewing their way across the Northern Range. In times past, many among the great herds would migrate out of the park every autumn, de-scending from the Yellowstone Plateau to the valleys beyond, where the win-ters ran shorter and the snowpacks shallower. Many of those timeworn paths had since been blocked by ranch fences and gauntlets of hunters lining up just across the border. The elk began bottling up inside the park. Willow, cot-tonwood, and aspen, the arid West's triumvirate of ecosystem-anchoring trees, began bearing the herd's full weight.

By the late 1920s, biologists were voicing concerns that critical browse plants were disappearing, that soils were eroding, that unpalatable grasses were proliferating. "The range," reported a team of scientists after visits in 1929 and 1933, "was in deplorable conditions when we first saw it, and its deterioration has been progressing steadily since then."

The park service responded by trapping and transplanting and—when those labors fell short—shooting Yellowstone's elk. Off and on for the next forty years the park's administrators culled the Yellowstone herds, though to no discernable improvement of the range. Occasionally there came voices from without suggesting a more holistic approach, involving a more expe-rienced class of hunter. Aldo Leopold obliquely broached this unmention-able in 1944, in a biting critique of the book *The Wolves of North America*. The book had ended on a disingenuous note from its coauthor Stanley P. Young, a career predator exterminator who suggested the country's few remaining wolves be allowed a few places "to continue their existence with little molestation"—at which Leopold fairly detonated:

> Yes, and so thinks every right-minded ecologist, but has the United States Fish and Wildlife Service no responsibility for implementing this thought be-fore it completes its job of extirpation? Where are these areas? Probably every reasonable ecologist will agree that some of them should live in the larger

national parks and wilderness areas; for instance, the Yellowstone and its adjacent national forests . . . Why, in the necessary process of extirpating
wolves from the livestock ranges of Wyoming and Montana, were not some
of the uninjured animals used to restock Yellowstone?

This was not a suggestion the managers of Yellowstone were racing to
embrace, given how proudly they'd just ridded their park of its predatory
vermin. Instead they redoubled their culling of elk, which lasted until the
late 1960s, when local hunters raised hell and their congressmen threatened to pull the plug on the park's funding. Yellowstone, in response,
adopted a politically expedient form of nonmanagement, marketed as "natural regulation." Old reports of elk damage were replaced with new ones
blaming the failing aspen on changing climate and fire frequencies. The
claims of overbrowsing and erosion were reexamined and declared exaggerations. The role of top predators was again dismissed as "a nonessential
adjunct to the regulatory process." Henceforth the naturally regulated elk
went on an extended bender, multiplying to stratospheric new densities,
and mowing down Yellowstone's woodlands.

In 1973, Congress intervened again in the park's ecological affairs, by a
far more roundabout route. That year, President Nixon signed the Endangered Species Act, a law calling for protecting and restoring species facing
extinction. The gray wolf of the Lower 48 was one of the first species named.
And when biologists started listing those few special places that might still
make a viable home for such a large, wide-ranging, pack-hunting carnivore
of hoofed animals, one particular piece of real estate—with its nineteen
thousand square miles of public land, centered by a national park bloated
with thirty-five thousand elk—invariably rose to the top.

On January 12, 1995, after nearly a decade of environmental impact
statements, court injunctions, raucous town meetings from Boise to Boze-
man to Cheyenne, and nearly two hundred thousand letters and public
comments, eight wolves—captured from the Canadian Rockies of Alberta
and caged in aluminum shipping crates—were escorted by a caravan of pa-
trol cars through the north gate of Yellowstone National Park. They rolled
past a roadside lined with cheering schoolchildren, past TV crews and

reporters, and onward out of sight, arriving at secret chain-link enclosures hidden in the hills above the Lamar Valley. Six more wolves arrived the following week and were taken to another pen above the Lamar. After a two-month acclimation period, the pens were opened. Within a week of leaving their pen, the Soda Butte pack had pulled down an elk calf. Wolves had come home to Yellowstone.

Biologists and wolf disciples of all stripes had long been waiting for the resurrection. The air above Yellowstone's Northern Range droned with the propeller of a Piper Super Cub, radio antenna drawing daily fixes on the wolves in their wanderings. Crews of wolf researchers cruised the winding highway of the Lamar Valley, tracking the electronic blips emanating from the collar of every wolf in the park. Amateur wolf-watchers gathered in flocks on the hillside, monstrous scopes piercing the distances to the daily spectacles on display in the valley. Yellowstone had instantly become the wolf-watching capital of the world.

One of the hints of the daily miracles to expect came the wolves' first year back in the park, when, with biologists watching and a movie camera rolling, two members of the Crystal Creek pack and a herd of elk put on a classic predator-prey clinic, from start to finish. The wolves' pace was at first unhurried—a loping exploratory gait that was later likened to "sifting through the elk as a shepherd would through sheep." One of the wolves found what it was looking for and sprinted headlong for a lone cow elk that was limping on a bad hind leg. The elk hobbled for the cover of the herd, but by then both wolves had singled her out. They cut the ailing elk from the herd, and the three took off across the valley flat.

A wolf is a long-legged endurance runner known for interspersing extended chases with sprints exceeding thirty miles per hour. And a healthy elk, as natural selection had decreed, tends to be yet a stride-length faster. But this was less than a healthy elk, and the wolves soon drew alongside, lunging for the neck. The elk shook them off, with forehooves flailing, trampling one of the attackers as she went. The wolf rolled and rejoined its packmate harrying the flanks, leaping and locking on with a bite force of sixteen hundred pounds per square inch. The elk faltered and fell, two wolves on her throat. She struggled to her feet, then fell again for the last time. Next day, biologists examined the kill, finding the elk's ankle arthritic and "swollen like a melon"—an inherited flaw perhaps, which had just been weeded from the gene pool.

It was a textbook enactment of the wolves choreographed dance of death with their prey—the casual scanning of the herd, the targeting and testing for weakness, the escalation of the chase, the flailing of hooves and locking of teeth on hide, and the wrestling to submission.

The wolves had rapidly assuaged the biologists' first concerns: They were feeding and mating and multiplying like mad, filling the predator void as fast as anyone had hoped or feared. Over the next ten years, the Yellowstone wolves spread to all corners of the park and spilled across the borders, south to the Grand Tetons and Wind Rivers, west to the Gallatin Canyon, east to the Absaroka Range, the population building to more than three hundred wolves across the Greater Yellowstone Ecosystem.

A feast of science, it would come to be called.

Hypotheses

What was known about wolves before Yellowstone had rested largely on a small handful of seminal studies, led by sturdy, woods-wise biologists toiling in lonely wilderness outposts in exchange for the rare glimpse of wolf society.

The first formal study of wolf and prey had been pioneered in Alaska, by the indomitable Adolph Murie. From 1939 to 1941, Murie hiked and climbed, dogsledded and skied thousands of miles through the vast tundra kingdom of Alaska's Mount McKinley National Park (now named Denali), all for whatever hints he could gather of the hunting lives of McKinley's wolves. The wolves' main prey was the Dall sheep, a snow-white subarctic cousin to the Rocky Mountain bighorn. For all Murie's Herculean labors, he observed but two live chases, neither ending in injury. But from what he could surmise from his crime-scene reconstructions of tracks and scat and blood in the snow, the wolves of McKinley worked hard for their living, chasing often and capturing seldom. The sheep, in their element among the rocky crags, sometimes toyed with their attackers. Murie once watched lambs a week old running circles around a wolf on a steep slope. He later found the sheep and the exhausted wolf lying fifty yards apart, the sheep dozing and resting and staring off in every direction.

In Murie's classic recounting of his Alaskan adventure, The Wolves of Mount McKinley, predator and prey emerge as ecological players and evolutionary partners, smashing old stereotypes of the wolf as rabid agent of destruction.

It is my impression that the wolves course over the hills in search of vulnerable animals. Many bands seem to be chased, given a trial, and if no advantage is gained or weak animals discovered, the wolves travel on to chase the bands until an advantage can be seized. The sheep may be vulnerable because of their poor physical condition, due to old age, disease, or winter hardships. Sheep in their first year also seem to be specially susceptible to the rigors of winter. The animals may be vulnerable because of the situation in which they are surprised. If discovered out on the flats the sheep may be overtaken before gaining safety in the cliffs. If weak animals were in the band, their speed and endurance would be less than that of the strong and they would naturally be the first victims . . .

The possibility is generally recognized that through predation the weak and diseased are eliminated, so that in the long run what seems so harmful may be beneficial to the species. Perhaps the evolution of the mountain sheep has progressed to a point where it is in equilibrium with its environment but still requires environmental stresses such as the wolf to maintain this equilibrium.

Murie's prescient observations in Alaska would echo thirty years later at an unplanned predator lab anchored fifteen miles off the northwest shore of Lake Superior. Isle Royale had been pioneered by moose around the turn of the twentieth century and by wolves fifty years later, the former supposedly swimming from the mainland, the latter crossing on a winter bridge of ice. In 1958, a Purdue University grad student named L. David Mech began what has since become the longest-running study of wolf and prey. What Murie had postulated about the wolves of Denali, Mech and a succession of students confirmed at Isle Royale: Wolves killed neither randomly nor easily.

Of seventy-seven moose that Mech saw tested by wolves, seventy-one walked away.

In time, Isle Royale also buttressed Murie's postulate that the wolf's impact extended beyond its immediate prey. Mech's most venerable successor, Rolf Peterson, and a colleague Brian McLaren measured the tree rings of balsam fir on Isle Royale, a prime winter moose food. The rings, layers of wood laid down during the growing season, measured wider in good years, thinner in lean years. If a hungry moose were to camp beside a balsam fir, the stress of being eaten would be gauged in the tree's razor thin growth ring the next season. When McLaren and Peterson compared the varying widths of the tree rings with the fluctuating number of wolves in the island, they uncovered an uncanny synchrony. More wolves, fewer moose, fatter rings, went the explanation. "When wolf numbers were high, the forest grew," Peterson exclaimed. "What an impressive achievement for a couple of dozen wolves that were just doing what comes naturally!"

From Isle Royale came the realization that the wolf was something more than a weeder of weak moose. It was a force of the forest. And that was a notion that Yellowstone, not long into its own experiment with wolves, was about to validate in spates.

A New Order
What Murie and Mech had learned through hard hours of slogging for murky glimpses through trees and postmortem reconstructions of ripening carcasses, Yellowstone not uncommonly offered up in living Technicolor from the roadside. With the throngs watching, the cameras rolling, and a naked clarity unmatched by any wolf terrain in the world, the wolves and elk of the Lamar acted out their parts—the predators methodically testing and selecting, chasing and attacking, sometimes killing, more often passing; the elk in their turn trotting and galloping, confronting and kicking, sometimes dying, more often escaping, and occasionally killing their would-be killers.

Much as the contests between wolf and elk dominated the highlight reels, no one creature so drastically reflected the wolves' return as the coyote. Better known in leaner environs as hunters of rodents and rabbits—more famously painted by Mark Twain as "a slim, sick and sorry-looking skeleton . . . a living, breathing allegory of want"—the coyotes of Yellowstone had developed into a different animal during the wolves' seven-decade absence. They had taken to forming big packs, killing elk calves, and occasionally, when emboldened by deep snow, tackling grown elk. They were seen stealing kills from mountain lions. Left alone in the land of giant prey, the coyotes had gone wolfish. This made for some interesting reunions when the lords of the house finally came home.

With the wolves' return, Yellowstone's coyotes were quickly reeducated on top-dog protocol. Those caught trespassing in wolf territories were chased down and torn apart. Within three years of the wolves' reoccupation, the coyote population of the Northern Range had been halved.

As dim as life had become for the demoted coyote, it had brightened for those the coyote had once terrorized. For the first time in more than a decade, the park's precarious herd of pronghorn antelope began to show signs of life.

The pronghorn, hoofed rocket of the West's open spaces, had been nearly annihilated by market hunters a century before. It had then made a celebrated comeback over much of its native range. Yet for puzzling reasons, the pronghorn had been suffering an unusually tough time surviving in the supposed sanctuary of Yellowstone. Their difficulties, it turned out, had much to do with the Yellowstone coyote. By observing pronghorns late in their pregnancy, the coyotes of Yellowstone had learned not only to take pronghorn fawns within hours or days of birth, but also their mothers as they defended them. In the first year that pronghorn biologist John Byers began watching the proceedings in the northwest corner of Yellowstone, seven mother pronghorn were killed within the first week of parturition. The average lifespan for the fawns was six days.

Meanwhile, eastward in the valley of the Lamar, life for the pronghorn was taking a good turn. In the thick of the densest concentration of wolves on the planet, the fawns of Lamar were surviving with baffling success. Byers's explanation was this: While the wolves of the Lamar were busy running down elk and trespassing coyotes, the pronghorn of the Lamar were busy raising fawns beneath their noses. At the end of his field season, Byers concluded, "It's likely wolves are going to be the single-most important force to save the pronghorn of Yellowstone."

Even for the wretched coyotes, so ingloriously dethroned, life in the harsh new kingdom of the wolf came with its rewards. It came with the spoils of war. Though the stomach of a wolf could hold twenty pounds of meat, there were seldom enough stomachs in a pack to finish off a seven-hundred-pound elk, let alone a bison bull at twice the mass. Leftovers soon became the lucrative new way of life in Yellowstone.

Coyotes only began to enumerate the beneficiaries. They and red foxes, American crows and bald eagles, black bears and grizzly bears, Clark's nut-crackers, gray jays, magpies, golden eagles and turkey vultures, and record gatherings of common ravens—all of them, plus fifty-seven varieties of beetle, were to be found making a good living on wolf leftovers. The carcass, with its attendant crowds of professional bone-pickers and carrion gour-mands, had become the most conspicuous billboard celebrating the wolves' return.

The Ecology of Fear

The wolves had brought to Yellowstone a perpetual windfall of carrion, a never-ending scavenger's ball. But from where Bob Beschta was standing, there in 1996 on the broken banks of the Lamar, it seemed a bit premature to celebrate the reawakening of an ecosystem whose arteries were coming undone. Beschta was a river man, and the Lamar by his measure was going downhill in a bad way. Back at Corvallis, he gave a seminar to his colleagues on what he'd seen in his first visit to the Northern Range of Yellowstone. He exhibited the shorn willows and aging cottonwoods, the missing cohort of seedlings and saplings that by all means ought to have been growing flush along its rivers. He reviewed the prevailing hypotheses, the idea that a chang-ing climate was somehow to blame; that wildfire, with its cleansing, restora-tive powers, had been suppressed for too long. He rejected these in favor of the least popular hypothesis in the park: that too many decades of too many elk in Yellowstone were pinching the blood supply of the park's biological diversity. When Beschta mentioned to his colleagues that aspen—whose forests ranked second only to riparian ecosystems as the lynchpin of biological diversity in the arid West—was also on the wane in Yellowstone, his colleague Bill Ripple sat up.

What the river was to Beschta, the aspen was to Ripple. As an amateur photographer, he particularly courted the aspen for "those cryptic yellows and golds and orange of the fall leaves, against that low autumn sun angle, and that backdrop of blue sky." As a landscape ecologist, he knew the aspen as one of the premier magnets of wildlife in the western mountains—forage of herbivores, shaded refuge of wildflowers, food and lodging of birds. Yellowstone should be the last place to be losing its aspen, thought Ripple, and the first place a scientist like himself should be asking why. Ripple turned to his graduate student Eric Larsen and said, "What do you say we go out there and solve this mystery?" As Ripple and Larsen began reconnoitering their new study area, their interviews took them south of Yellowstone, to the adjoining Grand Teton National Park. As the two walked into the visitor center, there hanging high on the wall was a poster portraying the subject of the day—a grove of aspens, with their creamy trunks rising from a fresh bed of snow. Standing right of center, staring forth with piercing amber eyes, was a big gray wolf.

Ripple again turned to his protégé Larsen, asking, "What about wolves protecting aspen?"

It was just a thought, an aside perhaps magnified now through the romantic haze of hindsight. And the wolf staring from the aspen, truth be told, had been mechanically inserted there by the artist's own hand. But either way, that poster would come to hang in Ripple's office. Over the following year, that image of the guardian wolf hovered in the mind, as the two scientists toiled through the legwork of backtracking the aspen's downfall.

Digging through the archives, Ripple and Larsen uncovered photos and aspen surveys dating to the early 1920s. They backed them with two summers in the field, aging live aspens where they stood. Once back in Oregon, they reconstructed a century's history of aspen in Yellowstone. Larsen then tallied the results and walked them into Ripple's office.

At first there was nothing but old news and dead-end leads: Indeed the aspens of the Northern Range had all but ceased reproducing after the

1920s. And nothing in the park's climate record or fire history suggested anything magical about that date. The elk, for their part, had been there all along, but only since the 1920s had their browsing brought Yellowstone's aspen to its knees.

And then, from out of the shapeless thicket of data, sprang an unmistakable figure. The 1920s, the last years of the Yellowstone aspen, were the last years of wolves in Yellowstone.

The hypothesis rapidly assembled itself. Aspen were dying not from any apparent lack of fire or rain but from a preponderance of elk, by way of too few wolves. In the wolves' absence, elk had begun eating nearly every unprotected shoot in the park. From where Ripple and Larsen stood, the aspen's demise was a food-chain reaction, a trickle-down demise tracing missing wolves to freeloading elk to missing aspen—in the ecological lexicon coined by Robert T. Paine, a trophic cascade.

In publishing their inaugural paper on Yellowstone's aspen demise, Ripple and Larsen recognized their wolf cascade as heavy on conjecture, a hypothesis woven out of aspen cores and old photos, eighty-year-old anecdotes and ecological theory. As senior investigator and upstart Yellowstone outsider, Ripple said he realized he was "going out on a limb—on tree rings, as it were." Yet the soft-spoken visitor from Oregon offered no apologies to the resident competition, whose fire- and climate-based theories had creaked for too long on far more rickety foundations. To Ripple's mind, the importance of aspen to the biological diversity of the West, and the appearance of wolves maintaining them, was a story too compelling to bury until proven beyond all doubt.

A year later, he and Larsen were back, this time with a meatier paper that had two top Yellowstone biologists added to the byline and a title leaving little doubt of their convictions: "Trophic Cascades Among Wolves, Elk, and Aspen on Yellowstone's Northern Range." Yes, agreed the authors, including Douglas Smith, leader of the wolf reintroduction program, aspen were indeed sprouting taller in certain streamsides and wet meadows, which happened to lie in the most heavily trafficked wolf territories of the Northern Range.

Only time would answer whether the suspected wolf effect would be enough to allow the few aberrant sprouts to become Yellowstone's first new aspen forest in eighty years. But Ripple wasn't waiting. In the summer of 2001, he rented a cabin just outside the park's northern boundary. He thought he might do a little photography, maybe some writing. Inevitably he ended up doing little but contemplating a landscape newly infused with the presence of wolf. For the next five weeks Ripple traveled up and down the valley of the Lamar, "just watching and looking and feeling it," he said, his search image now tuned to the willow shrubs that intermittently poked their stems above the stream banks, never forgetting that this was now something more than a particularly famous stretch of Rocky Mountain floodplain. This was now the territorial heart of the Druid Peak wolf pack.

With every trip through the valley, he found himself gravitating to the point where Soda Butte Creek fed into the Lamar River. There at the confluence, he was struck by the valley's most incongruent flush of greenery, of willow thickets towering twelve feet above bared banks. Somehow or other, the elk had ignored these particular patches.

Or—it occurred to Ripple as he sat gazing down from the overlook— what if the elk had more purposefully avoided this place? What if this point of land, in the eyes of an elk, was a death trap? What if the fertilizer ultimately responsible for these blossoming banks was a healthy dose of fear?

Ripple sent a message to his Oregon colleague Bob Beschta to come have a look, to listen to an idea he couldn't contain. Beschta, who'd been simmering on a low boil since his dispiriting visit in 1996, immediately set out. As he was driving up through the valley toward his meeting with Ripple, he came around the corner to the Soda Butte Creek confluence, to the scene of Ripple's amazement. There, where five years earlier Beschta himself had vowed to do something about the deplorable barrenness of the riverbank, were newborn thickets of willow.

Beschta spent the rest of the day being chaperoned through the Northern Range, visiting all the resurgent patches that Ripple had scouted in his retreat. Notice where they're growing, Ripple pointed out. Islands and gravel bars, steep banks and gullies—one could easily imagine the hazards posed to a fleeing elk. These curious little sanctuaries of willow were quite possibly the work of the wolf.

Beschta was slow to swallow the pill. "My feeling was, yeah, wolves are in the system, but what are they gonna do?" he said. "They can't eat enough elk to make a difference. All I could see was elk numbers. Thousands and thousands of elk. And if there are a few wolves in the system, what could they do?"

But there was no easily dismissing the willows, sprouting so conspicuously where elk—it could now be imagined—perhaps feared to browse. That summer, Beschta undertook his own study in the Lamar Valley, this time focusing on cottonwood trees. Cottonwoods, together with willow, formed the mainstays of the broad-leafed forest that typically cloaked the streams and rivers of the American West. And cottonwood, like aspen and willow in the Lamar Valley, had become scarce to the point of widespread concern. As Ripple had done with aspen, Beschta repeated with the cottonwoods of the Lamar. He counted tree rings, looked through old photos. The youngest cottonwoods of the Lamar were sixty years old. He checked for causes. Changes in weather, shifts in the stream channel, floods, and fires—none bore a pattern that could explain the missing generations. There was only one variable that Beschta could detect that coincided with the end of the cottonwood's line, and it had happened early in the century. "Holy smokes," exclaimed Beschta. "It's a wolf story."

In 2003, Ripple and Beschta went to press with their first Yellowstone collaboration, delving more deeply into the beginnings of the willow's and cottonwood's recovery and honing their case for fear as a dominant force. Introducing their "terrain fear factor," they suggested there were natural incongruities on the land that had kept the elk from dallying too long, which in turn had spared the saplings. There were gullies and terraces, islands and fingers of land jutting in the stream. These were the same places where the willows and cottonwoods had finally begun to show signs of recovery—places that an elk in a land of wolves did not want to be caught dead.

Fear so neatly answered the question of why thirty years of sporadic gunning by park rangers had not accomplished what a half decade and a hundred wolves apparently had. A few sporadic weeks of rifle fire left the riverside unguarded the rest of the year. But for elk in a neighborhood patrolled full time by wolves, every minute lounging in the river bottom involved a gamble of lethal stakes. It wasn't so much a change in elk numbers as it was a change in elk attitude.

The streamside forests, in modest little spurts, were reawakening, thanks to what appeared to be a healthy dose of fear. In one of the greening tributaries of the Lamar, a beaver colony had set up shop, bringing along its busy waterborne commerce of wildlife. It was the first beaver colony in many years on the Northern Range, and the first of nine more to come within the wolf's first decade back. It was as if a seventy-year winter had finally broken, giving way to springtime in Yellowstone.

Ripple and Beschta followed with a stream of reports, each building on the last, emboldened by the ever escaping willows of Yellowstone, and buttressed by a swelling literature of top-predator ecology. Fellow biologists from as far afield as Russia were reporting similar cascades triggered in the absence of top predators. Next door in Grand Teton National Park, where the wolf and grizzly and hunting human had been missing for many years, the moose densities had quintupled, and their gnawings had left the park with a battered riparian forest missing several of its native nesting songbirds—songbirds that soon returned as the first wolf packs ventured down from Yellowstone. Farther up the Rocky Mountain chain, from the Canadian national park of Banff came a report of willows, aspen, beavers, and songbirds more commonly frequenting areas more heavily stocked with wolves. From Jasper National Park, near Banff, came news suggesting the recovering wolves there were coinciding with a twenty-foot spurt of fresh aspens.

Ecologists and evolutionists since Darwin have recognized predation as one of the fundamental drivers of the diversity of life, a reaper of the weak, mother of invention, inspiration of poets. Wrote Robinson Jeffers,

> What but the wolf's tooth whittled so fine
> The fleet limbs of the antelope?

But death by fang only began to gauge the power of the predator's bite. The fear of those fangs had become an evolutionary force unto itself. An animal might survive a few failures at finding food or a mate, but as the behavioral ecologists Steven Lima and Lawrence Dill so wryly observed, "Few failures, however, are as unforgiving as the failure to avoid a predator: being killed greatly decreases future fitness."

The fear could be seen in the heron, fishing under the cover of dusk in anticipation of raptors by day; in the kangaroo rat, the deer mouse, the fruit bat—night foragers by nature—shunning moonlit evenings for fear of owls and fellow nocturnal predators; in salamanders and dragonflies, sparrows, marmots, and antelope, all shying away from richer feeding grounds for fear of predators lurking there.

Fear had even been detected in the psyche of the grasshopper. In a famous experiment conducted in a field in northeastern Connecticut, grasshoppers caged with enemy spiders ate less grass and lived fewer days. No big surprise there, except that the researchers had diabolically glued the spiders' fangs together, rendering them harmless. Without delivering a single bite, the spiders had triggered a trophic cascade.

Such was the ecology of fear that Ripple and Beschta believed they saw reviving the groves of Yellowstone. A herd of elk in the open could be a formidable quarry. But a gangly ungulate clambering through a streamside logjam, or floundering up a cutbank coated in ice, was an animal more likely to be rendered as meat. Ripple and Beschta would stand on sites where aspen were growing, look around, imagine what an elk might be thinking, and ask, "What is it that would make me feel uncomfortable here?"

"It's kind of like being a hiker in bear country in Alaska," said Beschta. "If you're in bear country, you very quickly find yourself walking across a landscape of fear. You get these sensitivities when you walk into willow bushes. It makes a world of difference. In open terrain I see bears half a mile away, and it's not a big deal. But when I start getting close to shrubs or start coming up over a sharp rise, man, I can tell you the hairs on the back of your neck go up, because you're not aware of what's coming over that ridge."

Yellowstone's newly instated ecology of fear was an intuitive, irresistible story. Wildlife magazines and major dailies across the country ran the alluring saga of the wolf as ecological archangel, sending ripples through the food web, restoring a pulse to Yellowstone's moribund valley of the Lamar.

The science was so sexy—for some, too sexy by half. It was one thing to suggest spiders guarding grass in a square meter of Connecticut pasture, and yet another to link wolves to willows in a two-million-acre ecosystem called Yellowstone. There were grumblings among the park's biologists that Ripple and Beschta had jumped the gun, had gone off too fast with their risk-based hypotheses. There were nonbelievers in the wolf's almighty ecological powers.

And so, with each new paper, Ripple and Beschta would dutifully reexamine the competing hypotheses, with boring predictability. "We were hard-pressed to find any climate signature that would even come close to trying to explain why willows would fall apart like they did in the Lamar Valley, all at the same time, and then start to grow again the last decade," said Beschta. "And to credit a large flood with creating the necessary conditions for willows to reproduce is almost ludicrous. All you had to do was put a fence around them and willows did just fine. Same with aspens. Put a fence around them and they do fine. I think to hang everything on a large flood is just ignoring the fact that every place anybody put up an exclosure, woody browse species did just fine. The inability of aspen, willow, and cottonwoods to grow above the browse level of elk for many decades, and now their journey to recovery, is all coincident with one fact: the removal of wolves, and now the recovery of wolves."

Yet the murmurs persisted. All of which was perfectly understandable, if not quite entirely on scientific terms. "Yellowstone is the oldest park in the world, and there are people who have been around here thirty, forty years," explained Doug Smith, lead wolf biologist and diplomatic park service employee. "They know an incredible amount of the park, a great deal of detail. So whenever you suggest an overarching paradigm, they find some detail that doesn't fit. They reject the new paradigm.

"Then some people," continued Smith, "I guess, are just ideologically opposed to the idea that this could be going on."

For some, there were just too many variables involved in Yellowstone's transformation to bet a professional reputation on the single, sexiest one. There was drought, there were water tables fluctuating, growing seasons lengthening. The elk herd itself was in flux, its numbers falling steadily

since before the wolves arrived. (A fact that Ripple and Beschta themselves eventually came to incorporate into their wolf-centric hypothesis
for Yellowstone's botanical resurgence.) On top of all was nearly a decade
of mild winters and subpar snowpacks to consider, allowing elk the luxury of foraging the grassy hillsides all season long, of never being driven
by starvation to the stream bottoms where the emergency rations of willow lay exposed. At the end of his presentations, Smith would often hear a
familiar challenge: Just give me one hard winter, and if the willows make
it through, I'll be a believer in the wolf hypothesis.

With the winter of 2005–6, the skeptics got their wish. It was a deep,
classic Yellowstone snowpack, a difficult time for elk turned deadly with abnormally high temperatures that turned the snow to concrete. As the elk
went hungry trying to paw through to the upland grass, they turned heavily for the bottomland willow. By logic of the hard-winter hypothesis, the
willows should have taken a beating, the elk's pain of hunger trumping
their fear of wolves. If such a blip of bad weather could crush Yellowstone's
recovering riparian, it would do likewise to the wolf hypothesis.

Among those watching, Bill Ripple was more than anxious for the answer. That summer he hurried out to see for himself.

He came back with a smile. "The willows are doing just fine," Ripple reported. "Actually, they're continuing to grow."

"I guess the point I'm making," said Duncan Patten, a veteran riparian
biologist of the Yellowstone ecosystem and prominent opponent of the wolf
bandwagon, "[is] there's a multitude of factors that control how willows
do. I'm not discounting wolves. But I don't think they're the bottom line.
They're part of the equation. At this point, put me on the side of the equation as the doubting Thomas. We really won't know for decades how this is
going to shake out."

Patten's doubts were echoed by a few who had been focusing their attentions on the predatory act itself, and wondering themselves if the fear
factor had indeed been overblown. These were not grasshoppers or chipmunks facing inescapable death in the clutches of a spider or raptor. These
were powerful, quarter-ton adversaries kicking and butting and otherwise
having much to say as to whether and when they would be wolf food.

Daniel MacNulty, a young researcher dedicated to deciphering the attacks, had been observing the elk as anything but helpless meat on the hoof. Tracking the wolf hunts by truck and on skis, camping in igloos for weeks at a time in Yellowstone's backcountry, poring over attack films frame by frame, MacNulty dissected what had come to be called the dance of death between wolf and elk. The dance was a finely choreographed ritual between attacker and defender, proceeding through a series of predictable steps, each one escalating in intensity, from the ear-pricking moment of detection to the final throttling gasp of death. Rarely, on average, did the dance follow to conclusion. The elk in their meetings with wolves had a way of immediately announcing who was boss.

An elk with speed to burn would throw its head high and break into an exaggerated trot, or stiffly bounce into a four-legged pogo gait called stotting. Either gait was far slower than galloping. Both were akin to an Olympic miler looking back on the field and skipping into a schoolgirl's hopscotch. "Chase me if you choose" was the heart-strong elk's message to the wolf, "but in the end, you will be the more tired and hungry for the effort."

Even to catch an elk was no guarantee of success. This was when wolves sometimes died on the receiving end of a well-placed kick or thrust of an antler. Which said nothing of the perils involved when wolves took to tackling bison. Although nine of every ten animals killed by wolves in Yellowstone were elk, a few wolf packs had taken to specializing on the most formidable prey in the land. After watching what a three-quarter-ton bison bull could do to a one-hundred-pound wolf, one had to wonder why. The battle of the decade was caught on film in Yellowstone's remote Pelican Valley, an epic struggle between an old bull bison cornered by deep snow and seven members of Mollie's pack. The bull was like a giant boulder, with a neck as thick as a pier piling. Its head amounted to a horned battering ram, its hooves packing power to splinter skulls and pulverize internal organs. Mollie's pack, for its part, was a hulking gang, composed of exceptionally large wolves that had chosen the kamikaze's life in the most brutal environment of Yellowstone. MacNulty was watching, and filmmaker Bob Landis's machine was rolling as the wolves attacked fore and aft, biting at the flanks, grabbing for the nose. Horns and hooves sent wolf bodies flying like rag dolls in a fight that lasted from dawn to dusk and on into the night. The light of the next morning revealed the old bull dead. And Mollie's pack wasn't faring much better. Three of the attackers were limping, and a fourth lay curled beneath a spruce, soon to die with a broken leg and untold internal injuries.

Over the first ten years of the wolves' return to Yellowstone, eight other wolves would die in predatory battle, seven from the antlers and hooves of elk, and the eighth in a losing battle with a moose. "But one thing that's underestimated, and I believe it's common," said Doug Smith, "is injuries. We don't have any good tally of injuries, but frequently we see wolves limping." For every five elk attacked by wolves in Yellowstone, four would win their contest. Fearful though it may have seemed to be prey in wolf country, one had to wonder—as MacNulty and his fellow skeptics wondered—whether the inverse might have better captured the essence of Yellowstone's wolfdom. Amid such fierce prey, a trophic cascade was no trivial matter for a wolf pack to trigger.

Smith himself had arrived in Yellowstone as a bit of a skeptic. He had been there from the start of Yellowstone's modern wolf era; he had helped carry the first wolves to the holding pens and cut the holes that turned them loose. He had come with thirteen years of experience on Isle Royale, training ground for some of the wolf's greatest students, gathering data on balsam fir and moose and the most famous wolf cascade on record. Along the way, about every five years or so, he had also seen a succession of pet theories tossed on Isle Royale's scrap heap. Wolves in control one year, a hard winter the next, a debilitating infestation of ticks after that. As much as Isle Royale had greased the skids in Smith's mind toward believing in a Yellowstone wolf cascade, it had also infused a measure of trepidation for the quick and appealing conclusion.

But as Smith tracked his wolves through Yellowstone, every week without fail for the next thirteen years, the wolves softened his doubts. "You begin to get an intuitive feel for what's going on," said Smith. "You follow wolves, you hike and ride and ski and fly, day in and day out in the same country they use. You begin to get these scientific hunches."

Smith's hunches increasingly, if cautiously, credited wolves as essential conspirators in Yellowstone's revolution. As the mortality statistics began pouring in from radio-collared elk the facts began to back the fear theory. The edges and incongruities of Yellowstone's landscape had indeed become particularly dangerous places in a land patrolled by wolves. Plotting the places where the elk had died over the years, the dots on the map clustered along the drainages. And that was a phenomenon that Smith, from his hours aloft in the Super Cub, could personally confirm.

"I've seen hunts on the flat and even terrain where the elk turn on the jets and outrun wolves, unless there's something wrong with them. But I've seen other times where the elk is beating the wolves, they hit a stream course, and the wolves close the gap. When you weigh five hundred pounds, you have to slow down more. And when wolves hit that incongruity, they slow down less. They close the gap. The elk is coming out of the other side of the drainage, and that's when the wolf gets its teeth into the hindquarters, and that's when the battle begins."

What Ripple and Beschta had surmised from reading thousands of browsed twigs and measuring saplings sprouting along the stream banks, Smith had come to understand by chasing the tails of the wolves themselves. By the end of his first decade with the wolves of Yellowstone, Smith was on record as one of the believers. As he would often say, "In the decades to come, wolves may prove no less fundamental to the life of Yellowstone than water is to the Everglades."

Whatever the homecoming wolves had done for Yellowstone, they had positively sparked the imagination. In those patches of sprouting willows and aspen and resurgent cottonwoods, conservationists envisioned an antidote to the long malaise of the western rangelands and waterways. They imagined trees taking root, sloughing banks taking hold, shade spreading from the flourishing groves over the water's edge. Fish and aquatic insects would return to the sheltered shallows, warblers to the willows. Beavers would come and build their dams, their ponds punctuating the hurried water, with fish and amphibians, muskrats and mink, following in the wake. They imagined the West's hoof-beaten oases reborn. And jump-starting the revival, if the preliminaries held true, was as simple—and as difficult—as plugging the missing predator into the picture.

"Biodiversity is the big deal, and sustaining it is huge," said Beschta. "It's something in my career I would never have guessed: that wolves would control the character of rivers. It's something I've worked on all my life—streams and rivers. And here it is this four-legged critter doing it. It's pretty amazing. Wolves controlling rivers."

NINE The Lions of Zion

BY THE END of the first decade of wolves in Yellowstone, aspens were sprouting by the thousands, willows were going gangbusters, and beaver were returning to secluded reaches of the Yellowstone watershed, all for the first time in at least half a century—all in suspicious synchrony with the return of Yellowstone's wolves ·

The shaky limb that Ripple and Beschta had ventured upon had grown sturdier with time. With every new announcement of their latest findings, Ripple and Beschta waited and listened for the professional rebuttals— some new evidence or critique, some out-of-the-blue broadside that would cripple their argument. And on they waited.

Waiting was an activity Ripple and Beschta had learned to practice on the move. The question, to their minds, was no longer whether wolves bolstered the diversity of life in Yellowstone, but what that result heralded for the rest of the country. *Canis lupus* had gone from nationwide resident to a half dozen pinpoints on the country map, leaving grand swaths of wolfless terrain.

Their curiosities led them far beyond wolves and willows. From the aspen groves and riverbeds of Yellowstone, Ripple and Beschta journeyed back to the historical roots of top-predator ecology. They found themselves tracing the literary trail from the heretics Hairston, Smith, and Slobodkin, whose green world hypothesis in 1960—suggesting a planet kept that way by predators—had touched off a half-century war of ecological worldviews. They passed through the Olympic Peninsula of Robert T. Paine, chronicler of the sea star *Pisaster* as keystone predator of the rocky shore; to the Aleutian kelp forests tended by James Estes's sea otters; to Michael Soulé and the bird-guardian coyotes of Southern California. They noted the urgent new warnings from Venezuela, where John Terborgh's predator-free archipelago was in the throes of ecological meltdown.

Along the way, they continually crossed paths with Aldo Leopold, the father of wildlife management, and latecomer champion of deer-eating carnivores. Long before Bob Paine had coined the term, the trophic cascade concept had become Leopold's cause célèbre in his crusade to save America's deer-bitten forests. He brought stories of sterility from the fangless woods of Germany, starvation from the deer yards of wolfless Wisconsin. In the early 1940s, he and several colleagues started pulling together the histories of more than one hundred ungulate populations across the United States, noting a recurring sequence of predator eradication followed by exploding populations of their prey and ruination of their range. Irruptions, Leopold called them. "We have found no record of a deer irruption in North American antedating the removal of deer predators," he wrote.

Conversely, wherever large carnivores had been suppressed, the forests and range had been preyed upon. "Thus the Yellowstone has lost its wolves and cougars, with the result that elk are ruining the flora, particularly on the winter range," wrote Leopold in *A Sand County Almanac*, fifty years before scientists counting tree rings confirmed the fact for themselves.

History abundantly confirms that Leopold, as poetic defender of predators, was largely ignored by those in charge. What it less often mentions is the meaner face that Leopold had once worn in life. This was the man who, in 1919, as a brash greenhorn forester in New Mexico and editor of his own newspaper, had written, "Good game laws well enforced will raise enough game either for sportsmen or for varmints, but not enough for both . . . It is most emphatically a reason for going out after the last lion scalp, and getting it."

This was the man who twenty-five years later had turned about-face, penning the century's most celebrated paean to top predators. In the essay

"Thinking like a Mountain," Leopold recalled himself as that young hair-trigger ranger in New Mexico, at the dying side of a bullet-riddled wolf whose young family he had just helped ambush.

> We reached the old wolf in time to watch a fierce green fire dying in her
> eyes. I realized then, and have known ever since, that there was something
> new to me in those eyes—something known only to her and to the moun-
> tain. I was young then, and full of trigger-itch. I thought that because fewer
> wolves meant more deer, that no wolves would mean hunters' paradise. But
> after seeing the green fire die, I sensed that neither the wolf nor the moun-
> tain agreed with such a view.

Kaibab

Of all the irruptions that Leopold heralded, one became his crusading ban-ner. It was a horrific tale of meteoric rise and smoldering crash. It was one he never actually visited until long after the rubble had cleared. Yet the leg-end of the Kaibab, the most infamous deer disaster in history, was a legend largely built by Leopold.

The Kaibab Plateau is an eleven-hundred-square-mile block of earth thrust high above the arid canyonlands of northern Arizona. It is an island wilderness, bounded northward by Utah desert, southward by the mile-deep plunge of the Grand Canyon. The Paiute people named the plateau Kaibab, a "mountain lying down." The Kaibab is a displaced piece of north-ern Montana in Arizona, where snowstorms are not unexpected in May, where the summers are famous for shady parklands of pine and green meadows and flowerings of rhapsodic verse: "We, who through successive summers have wandered through its forests and parks, have come to regard it as the most enchanting region it has ever been our privilege to visit," wrote the geologist C. E. Dutton in 1882. "There is a constant succession of parks and glades—dreamy avenues of grass and flowers winding between sylvan walls or spreading out into broad open meadows. From June until September there is a display of wild flowers quite beyond description."

There was also once a conspicuous herd of mule deer upon the Kaibab, for which the plateau was set aside as a game preserve in 1906. Henceforth, deer hunters were banned, and deer predators were exterminated. Over the next twenty-five years, some six thousand carnivores were trapped and gunned off the Kaibab, including eight hundred mountain lions and the last thirty of its wolves. The deer in turn went on a tear. Estimates of the Kaibab herd leaped from four thousand to one hundred thousand.

The Kaibab became country where tourists in their motorcars came at dusk to marvel at deer meadows packed by the hundreds, while naturalists viewing the same scene began counting down to catastrophe. As the browse lines crept higher up the trees, desperate deer started tiptoeing on hind legs for the last edible twig.

In his 1949 novel, *The Deer Stalker*, the western writer Zane Grey laid out as good a rendering of the classic Kaibab account as any historian since. Building to the irruptive climax, Grey's character Jim Evers—based on the hunter Uncle Jim Owens, who was credited with having shot six hundred mountain lions off the plateau—ironically sums up the ecological bunglings that have led the deer to the brink. "I've seen this deer herd grow from five to fifty thousand . . . Wal, killin' off the varmints, specially the cougars, has broken the balance of nature so far as these deer are concerned. Herds of deer, runnin' free, will never thrive whar the cougars have been killed off . . . They just eat up everythin'. An' now they're goin' to starve or die of disease."

On the eve of disaster, a plan is hatched. The scheme is to hire a posse of cowboys and a line of Indians to drive deer by the thousands off the plateau, down the walls of the Grand Canyon, across the roaring Colorado River, then back up the other side to richer pastures on the South Rim. Most incredibly of all, Grey is not making any of this up.

The plan was not fiction. Nor did the deer turn out to be as witless as cattle. They reversed through the drivers, turning cowboys into Keystone Cops, as they bounded back into the Kaibab forests from which they'd come.

The comedy ended there. The following two winters, without ever stepping foot near the canyon, the deer took the dive that many had feared. In the winters of 1924 and 1925 they starved en masse upon the plateau. Hunters were later called in to foreshorten the misery. Within six years, the Kaibab herd had fallen by eighty thousand.

The moral that emerged from the Kaibab spoke of the madness of eradicating predators as a means to more deer. As Evers drawls in *The Deer Stalker*, "Men cain' remove thet balance an' expect nature to correct it." Or, as expressed in the King's English by the British ecologist Charles S. Elton in his 1927 classic *Animal Ecology*, "Here it was clear that the absence of their usual enemies was disastrous to the deer . . . for the deer as a whole depend on them to preserve their optimum numbers and to prevent them from over-eating their food-supply."

Aldo Leopold was nowhere near when the Kaibab herd imploded, but he would not escape the tremors. At an international biological conference in Labrador in 1931, Leopold befriended the eminent Elton, and soon thereafter was subscribing to Elton's pyramidal view of life, the one founded on those many, many plants at its base and topped by those very few big carnivores at its apex. Leopold pondered the Eltonian pyramid, contemplating the consequences when "larger predators are lopped off the apex . . . The process of altering the pyramid for human occupation releases stored energy, and this often gives rise, during the pioneering period, to a deceptive exuberance of plant and animal life, both wild and tame. These releases of biotic capital tend to becloud or postpone the penalties of violence."

As Leopold began gathering the stats for his 1943 paper on deer irruptions, he came upon another fascinating pyramid of sorts, this one drawn by D. Irvin Rasmussen, a gifted young naturalist who for his dissertation had spent two summers absorbing the ecology of the Kaibab. Rasmussen in his write-up illustrated the deer debacle as a steep mountain, their numbers rising sharply following the extermination of cougars and wolves, their numbers then falling sharply toward the infamous crash that left one out of ten deer standing. Leopold lifted and embellished Rasmussen's graph for his irruptions paper. And for half a century thereafter, the Kaibab legend, as told by Leopold, was gospel. By the 1950s, rarely a biology textbook was published lacking the Kaibab disaster as a lesson in man's ham-handed tinkering with mother nature's sense of balance. Hairston, Smith, and Slobodkin used the Kaibab as one of the anchors in their green world argument. Rachel Carson's *Silent Spring*, the most influential environmental book of the

century, included the classic Kaibab story as proof of "the dire results of upsetting nature's own arrangements."

And nobody hammered the point more poetically than Aldo Leopold
himself. In his opus *A Sand County Almanac*, he rolled the Kaibab and all the
West's deer irruptions into one defining parable:

> I have lived to see state after state extirpate its wolves. I have watched the face
> of many a newly wolfless mountain, and seen the south-facing slopes wrin-
> kle with a maze of new deer trails. I have seen every edible bush and seedling
> browsed, first to anaemic desuetude, and then to death. I have seen every ed-
> ible tree defoliated to the height of a saddlehorn. Such a mountain looks as if
> someone had given God a new pruning shears, and forbidden Him all other
> exercise. In the end the starved bones of the hoped-for deer herd, dead of its
> own too-much, bleach with the bones of the dead sage, or molder under the
> high-lined junipers.
>
> I now suspect that just as a deer herd lives in mortal fear of its wolves, so
> does a mountain live in mortal fear of its deer.

Caughley

But in 1970 there came along a wildlife biologist from New Zealand named
Graeme Caughley to challenge Leopold and the Kaibab legend. Otherwise
known as the eminent authority on Australia's kangaroo population, Caughley was moreover an acerbic antagonist of shoddy science, whose idea of
fun was to skewer cherished paradigms, then to "sit back and say to yourself, 'Try to shoot that one down, you bastards.'"

"Charming," Caughley had once written of himself, "I am not."

When Caughley came upon the mountain that Leopold had built, he
took a bulldozer to it. That famous Matterhorn peak of Kaibab deer had
been constructed largely on loose estimates, Caughley chided. Leopold had
not only lifted Rasmussen's interpretation; he had then dramatized the
drawing by sharpening the downside of the deer peak from a steep slide to
a veritable cliff. The latter was the version that most textbook writers had

faithfully reproduced—a fact Caughley graphically displayed to everybody's embarrassment in the professional journal Ecology.

Then there was the underlying logic itself, that the missing predators were largely to blame for the irruption. What about the huge exodus of livestock— some two hundred thousand sheep, some twenty thousand cattle—that had coincided with the plateau's runaway population of deer? To Caughley's mind, that sudden surplus of forage was at least as good an explanation for the irrupting deer as that of a few less wolves and cougars.

"Data on the Kaibab deer herd in the period 1906-1939 are unreliable and inconsistent, and the factors that may have resulted in an upsurge of deer are hopelessly confounded," Caughley wrote in his 1970 critique. "The study is unlikely to teach us much about eruption of ungulate populations."

Boom.

It took a few years before the textbook writers and conservation historians caught up with Caughley's critique, but as word got out, Leopold's mountain was abandoned in droves. A raft of subsequent textbooks started expunging the romantic version of the Kaibab legend. Confessed professor C. John Burk, "I personally blench recalling instances—at least yearly for more than a decade—when I myself used, often dramatically and with gestures, the Kaibab example in a classroom situation . . . . One still cannot contemplate the Kaibab incident without extracting some moral from its consequences; all things considered, Caveat emptor would seem as good as any."

In time, even Leopold's sacred wolf-killing confessional took a bad tarnish. It was true that Leopold had begun his career preaching death to all varmints. And it was true that Leopold ended his career pleading for their salvation. But in between lay some shaky truths, not the least of which was the sacred wolf that Leopold had supposedly shot. The most famous martyr in conservation history might itself have been a myth. Leopold biographer Curt Meine could find no trace of it in the naturalist's meticulous hunting journals. Granting benefit of the doubt, Meine concluded that the wolf killing must have come during a period for which Leopold's diaries were confiscated, thereby placing the purported epiphany to sometime in 1909—

an unfortunate date, given it wasn't until 1941, after finally visiting the Kaibab, that Leopold completed his turnabout on killing predators.

And so it was that the Kaibab legend limped into the twenty-first century—just in time to be revived by a few wolves recently returned to northwest Wyoming. If the news coming out of Yellowstone could be believed—if a large terrestrial ecosystem could teeter upon the fate of an apex predator—why couldn't the Kaibab have happened as Leopold had described? In 2002, Dan Binkley of Colorado State University and three colleagues officially reopened the Kaibab case. And by the end of their second field season on the plateau, they were ready to see the textbooks rewritten once again.

Binkley and crew looked at aspen and deer on the Kaibab, as Ripple and crew had looked at aspen and elk in Yellowstone. And like their colleagues farther north, the Binkley team found the Kaibab aspen quitting production in approximate synchrony with the killing of the Kaibab predators and the irruption of Kaibab's deer.

One could almost hear Caughley objecting from the grave: What about those gazillion sheep grazing the Kaibab?

But this time it was the great critic's turn to be called on the carpet. Caughley had misinterpreted the historical records, badly overestimating the sheep numbers. "There were never that many livestock on the plateau itself," said Binkley, who double-checked the figures. "Livestock herbivory appears to explain none of the anomalies in aspen numbers."

In one serendipitous little sideline discovery, Binkley's crew dug up a 1935 edition of *Grand Canyon Nature Notes*, containing a clue to the missing aspen: "In marked contrast to . . . the Kaibab Forest, one finds rather extensive reproduction of aspen in the vicinity of the Quaking Aspen Ranger Station." It turned out a cabin had been homesteaded there in the mid-1920s by a Ms. E. Vaughn, whose family ran a few cattle, hunted cougars with dogs, and had built a corral. "Here a number of hounds are kept," read the report, "and of course, no deer come near."

Binkley's crew located the old corral, conveniently still flagged by the aforementioned grove of aspen. The scientists counted tree rings. The grove had sprouted soon after the fence went up in 1925, near the peak of the deer irruption. Outside the fence, no new aspen were to be seen for the next half century.

"Was Aldo Leopold Right About the Kaibab Deer Herd?" asked Binkley and his coauthors in a 2006 issue of the journal *Ecosystems*. Putting all the pieces back together—the hunting records of predators swept from the plateau, the tree rings revealing aspen soon thereafter eaten to a standstill by what must have been a tremendous herd of deer—they ended with the following: "We suggest . . . that the story (and the history of the story) be reinstated in ecology textbooks." The Kaibab legend breathed again.

Cougar Cascade

By the time the Binkley findings were heading to press, there was another study underway, just sixty miles north of the Kaibab in the canyon country of Utah, which was about to cast another strong vote Leopold's way. Its investigators were none other than Bill Ripple and Bob Beschta, who in tracing Leopold's countrywide register of deer irruptions, had come to the national park named Zion.

A monument not unlike the Grand Canyon's Kaibab, Zion is a ten-thousand-foot block of continental crust on the edge of the Colorado Plateau, long ago hoisted aloft and then cut from above by streams and great depths of time to form a dizzying land of kaleidoscopic sandstone cliffs and canyons. Zion was designated a national park in 1918, a mixed blessing that replaced a half century of homesteading in the bottomlands with a new culture of highways, foot trails, and tourists. By 1934, the yearly number of visitors had reached nearly seventy thousand, the bulk of them piling into the park's namesake gorge, Zion Canyon.

With the coming of the crowds went the cougars, and in the all-too-familiar sequence, the mule deer of Zion began amassing. The gallery forests began dying, the riverbanks began sloughing, and the deer took on a sickly pallor. As noted in a 1938 park service report: "The deer in Zion [canyon] are in very poor flesh and present a sorry spectacle. Professional wildlife visitors to the park this summer have been unfavorably impressed by the unbalanced deer forage situation here." By 1942, it was reported that "over-population of deer in Zion Canyon, with consequent scarcity of feed, is one of the major problems at present in Zion National Park . . . Vegetation is so over-browsed that it is in serious condition and there is danger of complete destruction." That year, the park service resorted to shipping deer from Zion.

This of course was familiar territory for Ripple and Beschta, with two critical and far-reaching distinctions. First of all, Zion's was a food pyramid topped not by coursing packs of wolves but by solitary, stalking cougars. *Felis concolor*—cougar, mountain lion, catamount, panther—is by any name the fourth largest cat in the world and pound for pound the ranking giant killer, leaping from ambush, fangs breaking necks and severing vital arteries, credited with killing mice to moose, one on one. And whereby elk were the staple of Yellowstone's wolves, mule deer were the standard fare of Zion's lions.

Zion National Park's second distinction stemmed from a fundamental and fortuitous twist of natural history. Over the ridge from Zion Canyon, only nine miles as the raven flew, flowed a nearly identical creek and corresponding gorge but with a major difference. While the commercialized Zion Canyon honked and thrummed with the annual commerce of three million sightseers, the adjacent canyon of North Creek remained roadless backcountry, its towering walls echoing silence but for the lonely footsteps of a few hardy hikers. In large part for these reasons, hardly a sign of mountain lion was to be found in Zion Canyon, while the streamside sands of North Creek were pocked with paw prints and scats. The sister canyons offered mirroring views of life, differing solely in the one variable of most concern to the scientist seeking the trophic cascade: the presence of lions.

These were not wolves, and this was not Yellowstone. If an ecological cascade were to be found here, triggered by such a vastly disparate predator as the cougars, in such starkly divergent environs as the canyon country of Utah, one would be left to ponder much broader concerns in the larger land of vanishing predators.

Escaping the bustle of Zion Canyon, Ripple and Beschta hiked over the ridge and down the canyon wall into the backcountry of North Creek. As the two began the protocol they'd so often practiced at Yellowstone—the dry recitation of stream depths and tree diameters, calling out numbers amid the burble of flowing water—it became apparent there was something more going on in North Creek than an inordinate growth of cottonwoods. The two could not help noticing the contrasts in life between the sister canyons. Streams of butterflies sailed by—swallowtails and sulphurs, blues and satyrs, monarchs and skippers. They had to look to keep from stepping on the cardinal flowers and red-spotted toads. The stream banks were a scurry of canyon treefrogs and spiny lizards. The waters shimmered with speckled dace and other native fish of the desert stream. North Creek flowed with a lifeblood that had been drained from the canyon of Zion.

The data sheets soon started filling with the names of flora and fauna. In addition to forty-seven times as many cottonwoods, there were three times as many lizards in North Creek, five times as many butterflies, more than one hundred times as many toads, more than two hundred times as many frogs. The disparity in cardinal flowers and asters grew to infinity, given the biologists couldn't find a single flowering stem growing along their transects in Zion Canyon.

Zion Canyon's trophic cascade took an entire page to illustrate. Beginning with the incursion of humans running the lions out of the canyon, it ran to the irruption of mule deer, to the death of cottonwoods, to the dissolution of streambanks, to the disappearance of wildflowers, and the butterflies that once sipped their nectar and spun cocoons in their leaves, and the sparseness of lizards that had once chased the butterflies and scampered for cover among the shrubs, and the rarity of frogs and toads and desert fish no longer courting in the pools that were no longer there.

From busloads of tourists to missing butterflies and bankrupt streambeds, Zion Canyon's cascade had ended in what the ecologists called a catastrophic regime shift. Zion minus its lions had lost more than cottonwoods in the bargain; it had lost a community. And if not for a standard of purity surviving over the next ridge in North Creek, where cougars still reigned, nobody would have ever known the difference.

Sixty miles south and sixty years before, Aldo Leopold may have stretched the facts in engraving the Kaibab legend. For that matter, he may never have even watched any fierce green fire dying in any wolf's eyes. But artistic license aside, he may have gotten the message right just the same.

As for Ripple and Beschta, never ones to accumulate moss, within the next two years they had blown the cover from two more supposedly sacrosanct refuges. In Jasper National Park in Canada and Wind Cave National Park in South Dakota they came to find the same disturbing pattern: Eradications of apex predators, dying trees, ecosystems sliding toward duller, less stable planes of simplicity. "Yes, we're finding cascades just about everywhere we look," said Ripple, who by then with Beschta was already on to another undisclosed site in their quest for ecological cascades.

It had become almost too easy to put a finger on the map and find communities there cascading away through the void of missing predators. It would be another matter to imagine a plan audacious enough to reassemble all the missing pieces.

TEN Dead Creatures Walking

IN SEPTEMBER 2004, a dozen biologists gathered at a ranch in southern New Mexico to kick around an idea that in the simplest of terms would raise the American landscape from the dead.

Their idea was to restore a certain character of life and wildness to what they saw as an ecologically eviscerated continent. They had seen for themselves the signs of sickness in the wake of the missing predators, in the waves of hoofed creatures and the withering forests and invading weeds. They had also seen signs of repair, in the reawakening of Yellowstone with wolves again at the top of the food chain. More to their point, they had come to realize grander and crazier possibilities. The fossilized bones of America revealed a land so recently rumbling with giants. It was a land of mammoths and five-ton sloths, herds of wild horses and bison with horns spreading six feet across. It was a risky terrain running with wolves and lions, with a cat bearing six-inch canines and a bear big enough to make a grizzly cower. To their imagining eyes, it was America at its full-blooded best. And their idea was to bring it back.

One might imagine them dreamers. Many would more easily consider them kooks. But this was not a dream, and their ranks consisted of some of the most respected conservation scientists in their field. This was not even a new idea. Some of them in fact had been seriously contemplating this for a long time.

Blitzkrieg

One wintry weekend outside Montreal in 1956, a young postgraduate ecologist found himself hunkered inside against the cold, pondering death. Paul S. Martin was preparing a seminar on the biology of the Pleistocene, an epoch beginning 1.8 million years ago with the onset of its signature glaciers, and ending—as Martin was now reminded—with the sudden and sweeping disappearance of so many great animals. Martin began the weekend thumbing through the time line of mammalian evolution leading up to the Pleistocene. With every advancing epoch, Miocene to Pliocene to Pleistocene, twenty-four million to five million to nearly two million years ago, the fauna had tumbled by the score over a cliff of extinction. Each die-off was followed with a resurrection of sorts, an evolutionary reconfiguration that restocked the empty niches with new and equally fantastic life-forms, large and small, a revolving canvas of bestiaries. And then, near the end of the Pleistocene, the pattern took a twist that brought a double take from Martin.

Again the reigning fauna had collapsed, but this time with a blatantly heavy bias. This time, the reaper had apparently handpicked the mightiest among the beasts. In contrast to the more random assortments that preceded them, the animals that disappeared at the end of the Pleistocene were almost without exception the largest of the lot. The cast had included Columbian mammoths, thirteen feet at the shoulder, and herds of wild horses and giant bison. America had housed giant camels and a beaver as big as a bear. There was everybody's favorite Paleolithic monster, *Smilodon*, the saber-tooth cat. There was a dire wolf and a rocket-fast feline called the American cheetah. There was a bear the shoulder-height of a moose with the leg-speed of a quarter horse, a creation to give anthropologists the night sweats.

And yet so suddenly they were all but gone, with only a skeleton crew surviving. Prior to that night in Montreal, Martin had once entertained notions of a career as an ornithologist. That would now have to wait for some other lifetime. The next year Martin moved his family to Tucson to take a research post at the University of Arizona's Desert Laboratory, where for the next fifty years he would work on solving the most contentious mass extinction since the dawn of humans.

The Pleistocene die-off in itself was nothing new to science. Charles Darwin had puzzled over it as a green, twenty-four-year-old naturalist circling the globe aboard the HMS *Beagle*. "It is impossible to reflect on the changed state of the American continent without astonishment," he wrote in his journal. "Formerly it must have swarmed with great monsters: now we find mere pygmies."

Alfred Russel Wallace, Darwin's codiscoverer of evolution by means of common descent, had too marveled at the great bestiary whose remains had been unearthed in the Pleistocene beds not only of North America but around the globe. "We live in a zoologically impoverished world, from which all the hugest, and fiercest, and strangest forms have recently disappeared," he wrote.

Surely, thought Wallace, in line with many of his contemporaries, the Pleistocene ice must have been the dagger that killed the megafauna. Yet even as he argued the case for climate's role, Wallace was subconsciously implicating a second party. "This is certainly not a great while ago, geologically; and it is *almost* certain that this great organic revolution, implying physical changes of such vast proportions . . . has taken place since man lived on the earth."

Since man lived on the earth. The point struck Martin a hammer blow. The glaciers had waxed and waned, climates had blown hot and cold with the wild swings of the Pleistocene, but the first arrival of humans on the continent had only happened once, and that, it appeared, had coincided all too suspiciously with the disappearance of the megafauna.

Martin's suspicions had come at an opportune time. The field of paleoecology had recently entered the revolution of radiocarbon dating, a technological timepiece with which one could precisely age life-forms long dead by measuring an odd form of carbon in their bones. Carbon 14 is a rare and radioactive molecule, a trillion times less common than its ubiquitous cousin carbon 12, but nonetheless taken up in measurable quantities in the tissue of all living matter. Carbon 14 is an unstable molecule, steadily leaking neutrons like grains of sand from an hourglass. When life stops, carbon 14 starts vacating its host, dwindling by half every 5,370 years. With a good sample, geochemists could date the death, to within a few centuries, of an animal that had died up to 40,000 years ago.

What the radiocarbon had begun to reveal was a North American menagerie of some forty species of mammal, all weighing more than one hundred pounds, coming to an abrupt end. All but fourteen had disappeared, and most of the casualties indeed clustered around the arrival of humans.

To those of the climatic persuasions, it was telling that 13,000 years ago—the cliff in time over which most of the megafauna of North America tumbled—the land was coming out of the latest glacial cycle, warming and drying and generally rearranging the continent's communities of plants. But to those leaning toward man as the smoking gun, it was more telling that most of these animals had already breezed through twenty-two such glacial cycles of the Pleistocene, some of them more severe than the latest.

Whatever or whomever the phantom killer was, it visited South America shortly after cleaning out North America, removing 80 percent of that continent's large mammals. And strangely enough, the same sort of blight had apparently blitzed Australia some 40,000 years ahead of the American massacre, taking the island continent's giant marsupial mammals, its giant flightless birds, and a lizard sixteen feet long. It did so too on the heels of human arrival, with no glaciers in sight.

In the Americas, the final years of the megafauna coincided with the first appearance of the Clovis culture and their exquisite spear points. Flaked from chert and flint and quartz, and sharp as broken glass, they turned up in brilliance and abundance in sites across the country. Some of them were recovered from between the ribs of fossil mammoths. The Clovis people, carrying the name of the dusty little town on the plains of eastern New Mexico where their artistic weapons were first uncovered, began showing up in North America around 13,400 years ago. They were the descendants of Siberian mammoth hunters who, during a period of low water, had walked across the Bering land bridge to Alaska. Southward they wandered, through an ice-free corridor, into a land of giant beasts that had never seen such odd creatures as these shifty bands of two-legged waifs wielding fire and hurling pointed sticks. To Martin, it was tantamount to a superpredator entering Eden. Spreading at a reasonable pace of 2 to 4 percent per year, cleaning out the naïve larder as they went, the Clovis colonists had in about two or three centuries laid waste to the continent's biggest mammals.

More damning evidence indicting humans as the Pleistocene asteroid followed in the boat wakes of ancient mariners. In their explorations of the oceanic islands of the globe, humans eventually discovered New Zealand and Madagascar, and again megafaunal carnage followed. The massive moas of New Zealand, the elephant birds of Madagascar, both families characterized by the big, the meaty, and the flightless, disappeared soon thereafter.

Whether by foul climates or foul play, the results of this pandemic of mass extinction were there for anyone to observe. In North America, the megafauna of the late Pleistocene had been extinguished by about 12,000 years ago. Which meant that by the time Columbus and company from Europe began their invasion 11,500 years later, the continent's fauna had already been thoroughly plundered. North America had been reduced to a tattered remnant of its Pleistocene grandeur. It had become Darwin's continent of pygmies.

Nor were the pygmies particularly safe. "When I consider," wrote Thoreau from mid-nineteenth-century New England, "that the nobler animals have been exterminated here—the cougar, panther, lynx, wolverine, wolf, bear, moose, deer, beaver, turkey, etc., etc.,—I cannot but feel as if I lived in a tamed, and, as it were, emasculated country . . . I should not like to think that some demigod had come before me and picked out some of the best of the stars."

The century following Thoreau amounted to a mop-up of the megabeasts. By midway through the twentieth century, the gray wolf and grizzly bear had been relegated to a class of loners, outlaws, and outcasts everywhere south of Alaska and the Canadian wilds. The secretive mountain lion survived in minimal numbers by keeping to the dark canyons and high crags, often operating under cover of night.

As far as Paul Martin was concerned, there was only one logical conclusion for the extinction. In 1967 he went public, in the magazine *Natural History*.

"Overkill," rather than "overchill," he declared, explained the missing megafauna. "My own hypothesis is that man, and man alone, was responsible for the unique wave of Late Pleistocene extinction."

And so the big gentle man with the impish eyes and matching streak of mischief became the prime minister of the blitzkrieg model, in which the Clovis big-game hunters, with their exquisite, deadly spear points, darted their way through the clueless megafauna of the New World, leaving behind the modern era's paltry cast of mammals.

But even as the quarreling paleoecologists chose their sides and dug their trenches, the leader of the overkill forces had already taken up a larger campaign. Martin had come to worry not just for the missing beasts but also for what they'd left behind. Throughout his homeland in the Southwest, the desert grasslands had for more than a century been retreating before a shrubland tide of mesquite and creosote, stems armed with thorns, leaves tainted with repellent tannins—an invading wave of defense specialists no longer thwarted by heavy browsers. Southward to Central America, fruits of tropical trees lay rotting on the ground, with no great elephantine mouths to swallow them, no giant guts to disperse their seeds in fertile piles of dung. The forests and plains with which the megabeasts had coevolved were lonely for the loss. So too were their few Pleistocene survivors. In the coastal mountains of Southern California, the last wild breeding pair of California condors—the last of the megafauna's specialist scavengers—were to be captured in the mid-1980s and taken inside for their own good.

"We live in a continent of ghosts," Martin would say. "Those who ignore the giant ground sloths, native horses, and sabertooth cats in their version of outdoor America sell the place short."

Martin had moved beyond the question of who done it, to the suggestion that something ought to be done about it. With almost every overkill paper and presentation from 1969 onward, he began inserting a familiar closing. His bedtime story could not end so bleakly, with humans extinguishing the megafauna. It had to continue with humans bringing them back. Martin lobbied for camels and horses as Pleistocene stand-ins to repatriate the Southwest scrublands. But for about thirty years, with the warring troops of overkill and overchill so busy lobbing fossil bones and radiocarbon bombs back and forth, nobody much heeded Martin's more heretical punch lines.

Rewilding

In the fall of 1998, Michael Soulé and fellow conservation biologist Reed Noss proposed in the journal Wild Earth an idea that had come to be called "rewilding." It was an idea prompted by several sobering observations on the state of modern conservation. One was that the most glorified national parks and refuges in the land, the supposed bastions of biological complexity, weren't cutting it. Just as the tenets of island biogeography had predicted, the isolated parks—besieged at their borders by foreign species and disease, weakened from within by inbreeding among the natives, and hazardously exposed to every natural cataclysm visited upon their lonely little fortress—were being drained of wild species at rates inversely related to their size. And with few exceptions, their size was far too small.

The other ubiquitous pitfall of the modern park, noted Soulé and Noss, was its conspicuous lack of big carnivores, itself a guarantee of attrition. Without Terborgh's "big things that run the world," the parks were ultimately left to the mercy of the herbivores and destined for ecological decay.

The answer to the parks' dilemma was a strategy niftily nicknamed the three C's: cores, corridors, and carnivores. Protect the biggest remaining pieces of nature, multiply their effective size by connecting them one to the other, and replace their missing pinnacles—put back their missing grizzlies, wolves, mountain lions, jaguars, wolverines, or whatever top predator no longer hunted there.

This was Soulé and Noss's vision of rewilding, at once praised for its soaring spirit and scientific validity and drubbed for its naïve impracticality. (In addition to restocking the country with big dangerous carnivores, Noss had earlier called for at least half the land area of the Lower 48 to be included in this network.) And yet, within a year, the idea of rewilding, as Soulé and Noss imagined it, was to be dwarfed in its audacity.

In 1999, also in Wild Earth, Paul Martin and a like-minded colleague named David A. Burney began by politely praising Soulé and Noss's laudable ambitions. Upon which, they pushed the plunger. "Here we consider the ultimate in rewilding . . . We suggest that the project begin by restarting the evolution of the most influential of the missing species, the extinct animals most likely to have exerted the greatest influence on their natural environment." There was no pussyfooting around what the two had in mind: The title of Martin and Burney's paper blared, "Bring Back the Elephants!"

If anyone might have doubted their sincerity, a year later Martin retested his and Burney's elephant bomb, this time at the annual gathering of the Society of Conservation Biology in Missoula, Montana. "In the face of a radical depletion of America's ice age megafauna we envision proactive experiments in restoration," he said. Bison, wild horses, and camels were ready and waiting for repatriation, Martin declared. The Columbian mammoth will never again browse the western streamsides, but why not its surviving kin, the Asian elephant? How about identifying a network of large parks or ranchers sympathetic to the cause? Never mind our outdated notions of the pure and pristine. Let's "restart megafaunal evolution in this hemisphere, and educate our children about its grandeur," Martin admonished.

After three decades on the rewilding stump, Martin by then had become used to the nods from colleagues who would politely clap while privately muttering away the idea as insane. This time, though, when Martin finished, he was approached by a sincere young man with light in his eyes. Josh Donlan, a recent graduate from the University of California at Santa Cruz, had already established a career of venturing where few in the mainstream of conservation dared go.

On oceanic islands the world over, where naïve island faunas were being obliterated by shipborne invaders in the form of feral cats and rats and livestock, Donlan and his colleagues had been devoting themselves to saving such island fauna—by annihilating their killers. Donlan in his young career had taken part in shooting more than 160,000 goats in the Galápagos to stop its famous tortoises from being run off the islands. In the Channel Islands of California, he and his cohorts had come to the rescue of a tiny seabird called Xantus's murrelet—to the howls and lawsuits of animal rights activists—by poisoning the island's entire population of ship rats that had been plundering the helpless birds where they nested. When Donlan happened into the auditorium to hear Paul Martin talking about putting Asian elephants in New Mexico, he had to chuckle at his own rodentine hurdles. He thought, "I got it made."

Donlan and Martin, conservation's wunderkind and sage of political incorrectness, began a conversation bound to lead to bigger trouble. Their first paper together took on the status quo of their profession's most sacred charm, the pristine myth of the year 1492. For so many in the field of conservation, Columbus's arrival in the New World had long stood as the baseline of ecological restoration—the beginning of the demise, Eden on the brink. By Donlan and Martin's measure, 1492 was a hollow target, a faunal façade masking a mammalian heritage already robbed of two thirds of its lineage. If restoration was honestly the goal, why not aim for the pinnacle of the Pleistocene? "The deep history of North America is commonly ignored in conservation strategies," they wrote. "In the process of returning the California Condor (Gymnogyps californianus) to the Grand Canyon, should we also return the kinds of animals the birds once fed on?" That is, if condors, why not also horses and camels, mountain goats and elephants?

Donlan ended up attending Cornell University, under the advisory of the evolutionary biologist Harry Greene, a serendipitous union that added yet another charge to the Donlan-Martin duo. It happened that Greene was a longtime friend and kindred soul of Martin's, a fellow admirer of venomous snakes, an old-school naturalist, and a disciple of biological diversity for beauty's sake.

Nearly twenty years earlier, Greene had been among the masses in his profession who gravitated to a new book edited by Michael Soulé and Bruce Wilcox, named simply Conservation Biology. It contained a chapter in which Soulé himself took a disheartening view of conservation's prospects. The tropics—the habitat that held nine of every ten species on Earth— were being slashed and burned at a hurtling pace. To Soulé it seemed the reserves were too small, the tropical floras and faunas too isolated, and the pace of extinctions too rapid for any new generation of species to rise from the ashes. "Thus, it would appear that with or without management, evolution of large, terrestrial organisms in the fragmenting tropics is all but over," Soulé wrote. Or, as he and his coeditor, Bruce Wilcox, would state more wistfully in the same volume, "Death is one thing—an end to birth is something else."

It was the worst scenario Greene had ever imagined. And yet here, almost two decades later, in the elephantine musings of his old buddy Paul Martin, in the unjaded optimism of his new protégé Josh Donlan, was a stab at redemption. To hear Martin and Donlan talk, one began to wonder: If elephants, camels, and wild horses still lived, at least somewhere in the world, why couldn't they live again on the continent that had birthed them? Why couldn't the descendants of the American lion, now surviving so tenuously in Africa, be allowed to chase them on its ancestral homelands?

The immediate answer, to most people, hardly needed stating. "Most people dismissed it as silliness," said Greene. "But the more we talked about it, Josh and I decided it's not silly. Let's put together a working group. Let's thrash it out."

The two assembled an eclectic team of twelve who gathered in 2004 for a long weekend at Ted Turner's Ladder Ranch in the Chihuahuan Desert of New Mexico. Among them was, of course, the patron saint of overkill, Paul Martin, and David Burney, the fellow farsighted paleoecologist who'd seconded their irreverent paper "Bring Back the Elephants!" There too was Michael Soulé, one of the spearheads of the modern discipline of conservation biology. The roll call also included Jim Estes, chief herald of the sea otter as marine ecology's classic keystone predator; Felisa Smith, an expert on Late Pleistocene mammal communities of North America; Dave Foreman, former congressional lobbyist and recent founder of the Rewilding Institute, a think tank for restoring large carnivores to vacant niches of North America; Joel Berger, innovative expert on large mammal conservation, who'd once sawed horns off African rhinos to spare them from poachers; Gary Roemer, a community ecologist from New Mexico State University; and two eminent biologists from Arizona, Jane and Carl Bock, who were invited to the meeting as a balancing voice of caution.

Over easels and PowerPoint presentations and after-hour beers, the Ladder Dozen dissected the rewilding idea, breaking it down to its factual nuts and bolts, its practical challenges and criticisms, its societal costs and benefits.

They agreed on a handful of otherwise unpopular premises: that human influence had utterly pervaded the planet; that conservation's prevailing benchmarks of purity, the 1492 landing of Columbus and the 1804 trek of Lewis and Clark, were in fact discoveries of an impoverished continent already plundered of its greatest predators and prey. Those benchmarks, at best, were rickety foundations on which to begin rebuilding the continent's native wildness. Why not raise the standard, to that more glorious and decisive moment some thirteen thousand years ago when people first set foot in North America, when the ecology of the continent was pulsing with whole blood?

The rewilders further agreed that the large animals' absence was to be ignored at great peril. The fossil record was unequivocal: For much of life's history preceding the arrival of the superpredator ape, life with megafauna had been the global norm. Forests, grasslands, and savannas had evolved in step with the Pleistocene megafauna. Their soils had been turned by trampling hooves, their seeds widely ferried and judiciously fertilized in herbivore dung. And for as long as there had been great plant-eaters, there had been great meat-eaters among them, checking their excesses, penalizing the slow and the careless.

More recent reports from the field were already previewing the consequences of the losses. All attending were familiar with the classic case of the kelp-saving sea otter, compliments of their colleague Jim Estes. Frightening reports of further ecological meltdown continued streaming north from John Terborgh's predator-free forests in Venezuela.

And there were yet more subtle cascades underway in the psychological realm. Joel Berger had spent most of his professional life studying the behavior of large mammals, from bighorns of the American Great Basin to black rhinos in the brushlands of Namibia. As an authority on big-prey species, Berger had taken a particular interest in the worldwide eradication of their predators. More specifically, Berger had evidence that even short-term absences of big predators could be deadly.

Berger was headquartered on the flanks of the Grand Teton range in Wyoming, where for more than half a century—since the grizzlies and wolves were driven out—the moose and elk had been free of serious predators. He was there waiting and watching when, with the turn of the twenty-first century, a few grizzlies and wolves from their recovering refuge of Yellowstone started roaming back into the Tetons.

Berger had an idea there would be trouble. He wasn't looking for dramatic chase scenes or bloody battles. He was looking for those subtle sirens from the moose's fight-or-flight alarm center—the twitch of a mane, the flare of a nostril, the raising of neck hair. These were visual cues requiring an unusual degree of intimacy between subject and observer—in Berger's case a potentially dangerous level of proximity. To observe such levels of detail would require Berger to mingle with the moose. So to better mingle with the moose, Berger became one.

Along with his research colleague and wife at the time, Carol Cunningham, Berger outfitted himself with a two-piece moose costume. Berger and Cunningham, head and tail, would sidle up to their subjects, concealing within their Trojan moose a multisensory armory of predator sounds and scents. They came armed with recordings of wolf howls; they carried snowballs soaked in wolf urine and authentic dungballs of bear. Lumbering within range, Berger—with the arm of a former semiprofessional baseball player—would toss the scent bombs under the noses of their subjects, and watch.

When the first dung started flying, the authentic moose just stood there. When Berger broadcast the howls of the wolf, the moose hardly stirred. Rendered naïve by too much time in a land stripped of danger, the moose had forgotten the smell of the enemy, the alarm call of the wild. After two hundred thousand years of the moose keeping one step ahead of their lethal enemies, just a few generations apart had closed the gap.

Berger later watched as wolves walked up on clueless mothers and carried off their calves.

To the moose's good fortune, the trauma of losing an offspring tended to reawaken dormant fears. To the sounds and smells of wolves, the ears began to twitch, the nostrils began to quiver, the hair on the mane stood on end. "Now they worry plenty," reported Berger. "When they hear a wolf howl, they run."

Contrasting all the bad news from the world of vanishing predators was the brightening beacon of Yellowstone. Blossoming under the new watch of the wolf, Yellowstone was reminding many of what a landscape had to gain from rewilding—as well as the penalties it suffered for withholding. By the rewilders' estimation, to risk attempting nothing on behalf of the missing beasts was to risk a depauperate world of weeds.

Which brought the Ladder Dozen to the rather imposing quandary of resuscitating a countrywide graveyard of deceased species. Their answer was, in a word, proxies—close relatives and ecological equivalents that would serve as megafaunal stand-ins and rekindle the evolutionary embers. The country was already well stocked with potential candidates. Not too far from where the rewilders were sitting, in the Hill Country of Texas, some seventy-seven thousand large exotic mammals were already roaming within the expansive confines of game ranches—among them camels, cheetahs, and myriad species of African antelope. For that matter, even bygone mammoths and mastodons were available for repatriation, if one were to consider their elephantine cousins caged in zoos across the country.

Here, by proxy, was a means of not only restoring North America's megafauna but also providing a fail-safe for endangered megafauna of the world. Wild Bactrian camels, on the verge of extinction in their last holdout in the Gobi desert, might find paradise in the prickly scrublands of the Southwest. Representatives of African and Asian elephants, both being slaughtered or swept aside in their home countries, could be protected too within American preserves. Here, finally, was a way to reawaken the sleeping predator, wherein cage-bound cheetahs and languishing lions might be turned out in fenced expanses to once again hone speed and wits in open pursuit of North America's repatriated hoofstock.

And there were obvious places where one could start to work: the vast open stretches of Great Plains steppe and Southwest scrubland. Though typically such country came laced with barbed wire and peopled with bad attitudes toward big carnivores, there was a good bet the prospects of tourist dollars funneling to otherwise dying rural economies might adjust some attitudes accordingly. If all went well with the trial runs, perhaps one day the fences could be moved back to accommodate grander arenas—Pleistocene parks, they would be called—in the widest unpeopled spaces of western

America. One could imagine elephants crushing creosote in the Chihuahuan Desert of New Mexico, lions stalking wild horses through Colorado short-grass, cheetahs chasing pronghorn through the Red Desert of Wyoming.

Such was the essence of the rewilders' ultimate vision. It had fences and fail-safes and experimental designs, it had humble admissions of problems yet unsolved, and it had hope. And by that Sunday night at Ladder Ranch, it had a draft with twelve signatories.

The stately journal *Nature* accepted the rewilders daring post, albeit a version drastically slashed and squeezed to fit as a two-page commentary. As the top two of the paper's listed coauthors, Donlan and Greene would naturally be the first in line to receive any public inquiries. Though both were battle-savvy veterans of the publishing process, neither was prepared for what was to come. Jim Estes, still freshly bruised from the stoning he received after suggesting that whalers might have triggered an ecological collapse in the North Pacific, had to wince over the drubbing he saw coming to his rewilding friends and coauthors from Cornell. "This whaling thing is small potatoes compared to rewilding," he warned them. "When this thing hits the street, you guys better put on your flak jackets."

Word went out, and word quickly came back, flooding Greene and Don-lan's e-mail boxes with vitriol. News bureaus on both sides of the Atlantic swooped in, smelling blood. Amid a few tepid nods of approval from adventurous ranchers and high-spirited laypublic, the jeers resounded.

"Impossible."

"Pure fantasy."

"A terrible and absurd idea."

African patriots savaged the American rewilders for their imperialistic gall. One wrote, "Thank you for planning to rob Africa of her animals so you can beautify your barren great plains with her native animals!! Why come after the animals that support our tourism. Leave Africa and what is hers alone, thief!!!!!!!!" (This despite the authors having mentioned five times in the space of two pages their intentions to use captive animals already living in America.)

Some in their imaginations painted Donlan and Greene as pointy-headed, ivory-tower idealists. Greene, a former army ambulance driver and impassioned handler of lethal vipers, and Donlan, whose field résumé included firing half a million rounds of live ammunition in the name of conservation, could only blink in astonishment.

Academics, some with offices just down the hall from Greene's and Donlan's, ganged up to chide their colleagues in print. Fourteen conservation biologists, including three senior scientists from the Nature Conservancy, signed on to a letter in *Nature* declaring the rewilding vision the wrong vision. Four others published a paper rating the rewilding proposal as only slightly less sensational than *Jurassic Park*, a fantasy novel and film in which sixty-five-million-year-old dinosaurs are brought back to life and begin eating people.

"I'm shocked that that even made it into print," said Donlan. "I mean, how many times does ten thousand years go into sixty-five million?"

Added Greene, "It's not a stretch to say that they mostly thought we were going to come dump a bunch of elephants on the suburbs of Topeka."

Amid all the scoffings and scoldings, Donlan and Greene were struck by the eerie silence in response to their central challenge. Nobody had addressed their premise, in which the absence of big predators and prey had lit an ecological and aesthetic time bomb. "That to me is the disappointing thing," said Donlan. "The papers that have come out so far are to some extent ridiculous. It's clear that there are huge challenges and huge obstacles to this idea, but in my opinion they haven't been discussed. Everyone's dancing around outside of this main issue."

It was particularly suspicious that nobody had made noise concerning previous like-minded efforts to restore butterflies, birds, and tortoises to places they no longer or had never lived. Nobody seemed concerned about northern Siberia, where the Russian ecologist Sergey Zimov was by then already deep into a Pleistocene rewilding experiment of his own, returning wild horses and bison and musk oxen to the bereft tundra. Nobody made a peep over one of the rewilders' own featured candidates, a Pleistocene relic the size of a coffee table named the Bolson tortoise, rescued from a holdout in Mexico and already thumping about the Ladder Ranch.

In the end it was the idea of the big and the dangerous returning to America's outskirts that had primed the explosion. One letter writer declared, "If an Elephant ever comes tromping through my yard it'll get an ass full of buckshot."

But ultimately, it was the lion that lit the powder keg:

"You are a fucking moron if you release killers in our homeland. I hope the cattle rancher guys shoot your ass or feed you to those lions if you release those killers into our ecosystem."

"I know we'll all appreciate it when are [sic] kids are eaten alive at a campsite, or when we get gobbled up while taking a hike."

"If they get near me, my family, friends or my property then I'll be 'really careful' when I place the crosshairs on them and slowly squeeze the trigger of my Remington 300 UltraMag. Are you sane?"

ELEVEN The Loneliest Predator

A risk-free world is a very dull world, one from which we are apt to learn little of consequence.

—Geerat Vermeij

In a quieter moment, two years after inciting international hysteria with their notions of repatriating North America's missing megafauna, Josh Donlan and Harry Greene sat down to reflect. In recalling the response—as one might try recalling the driver of the bus that had flattened them—they sifted through the blitzkrieg's rubble of radio spots, newspaper editorials, *Jumanji* cartoons and *Jurassic Park* jokes, the smoldering stacks of e-mail and whisperings in the hallways. Whatever noble message they might have intended, a far more menacing one had ultimately been received. The masses, skimming all the pros and cons and scientific protocols that the rewilders had painstakingly weaved into their message of revival and hope, had slashed and twisted the entirety to a sound bite of reply: No lions in my backyard. Donlan and Greene had at least identified the driver of the bus, and its name was fear.

For all the care given to laying out the ecological case for the big and dangerous beasts, the rewilders had underestimated the power the great predators still wielded over their own particular species. Rewilding had challenged the topmost survivor among the megafauna to consider lightening up its forty-thousand-year death grip. It had reopened ancient and frightening territory, in minds that subconsciously were still walking the plains naked among Pleistocene lions.

The Meat of Man

Some two million years before the first utterance of rewilding, a clan of apes swung down from the branches on the sere edge of the East African savanna and set out, in fits and starts, on foot across the open plain. They had unimpressive fangs and nothing in the way of claws one could call weapons. Their leg speed was laughable. Yet by venturing into the open, they were venturing into competition with an accomplished cast of the fleetest and most fearsome carnivores on the planet.

Out among a bristling menagerie of lions and hyenas, saber-toothed cats, wild dogs, and leopards, their forays might be construed as those of a fool or thrillseeker, some ancient progenitor of the modern rodeo clown. The walking apes were in fact pioneers of necessity. They had been pitched headlong into the Pleistocene's turbulent era of shrinking woodlands and spreading grasslands, their sanctuaries of trees receding into isolated pockets. Like tadpoles in an evaporating puddle, they either had to grow legs or lay exposed to the mercy of the elements.

Lacking brawn, the apes needed pluck and hustle to succeed at what would become a pirating life in plain view. Hunting for meat and marrow out on the naked spaces, the odd, two-legged newcomers learned to scan the horizons for descending vultures, to listen for the roar of lions, to track spoor on the run. If they could not subdue it, they would scavenge it. Competing with jackals and hyenas and vultures, veterans of the trade, they raced to reach the carcass first. And with time and hunger and a few tricks borrowed from the competition, they eventually learned to bully and bluff their way to the prize.

Somehow these sorry contenders had made camp in the gladiator's arena and lived to brag about it. The descendants of the skinny, odd simian would come to number six and a half billion people, swarming over nearly every foothold of terrestrial habitat on Earth, to the highest peaks and farthest poles and deepest oceanic trenches. They would eventually wield the power to level mountains, to dam the biggest rivers, to coat entire continents in concrete and crops, to alter the climate as it had once altered them. And before they were through, they would vanquish their old predatorial contenders to zoos and artificial reserves, to the slums and refugee camps and Siberian outposts of nature.

Some would mark the defining moment of human history by the first domestications of grains and beasts some ten thousand years ago. Yet the agricultural revolution was not some sudden blossoming of inspired genius but rather the inevitable expression of creative talents honed over the long haul of the Pleistocene. They were talents honed in the heat of competition with that cast of fierce predators, which not only challenged the upright ape to keep itself fed, but to keep it from becoming the food.

In 1924, in a cave on the edge of the Kalahari Desert in South Africa, a miner named de Bruyn discovered a skull that was soon hailed as the missing link. It had once belonged to a small, round-headed creature with little teeth and big, forward-peering, lively eyes—a child, three to four years old, weighing maybe twenty-five pounds, with a brain as big as a large gorilla's. Formally called *Australopithecus africanus*—the Southern ape—the child was nicknamed Taung, after a nearby railway station. And the paleontologist who chaperoned the Taung child into stardom, a spirited South African named Raymond Dart, declared that this immediate ancestor to the human line was a skull-smashing, blood-spilling killer supreme.

Dart, whose exploits were later glamorized by New York playwright Robert Ardrey in his book *African Genesis*, had examined the Taung child along with other early australopithecine man-creatures, in the context of the caves where they had come to rest. Their surroundings, eventually dated to more than two million years ago, had sealed the remains of thousands of other creatures who had visited or been dragged into the cave in their time. There were bones of antelope, baboons, hippopotamuses, giraffes, porcupines, jackals, and hyenas.

Most critically, to Dart's eyes, was the unusual preponderance of antelope leg bones, their knobby knees the likeness of a club. Nearby lay the dented skulls of baboons, and hence, by Dart's elementary deduction, the marks of violent death at the weapon-wielding hands of these vicious little ape-men. Dart's hypothesis of murder and mayhem at the root of man's ancestry, dripping of Ardrey's melodrama, captivated a huge lay audience.

It drew a cooler reception from Dart's colleagues, who found it hard to swallow the incongruence of this supposed two-legged terror. This, after all, was a line of man-apes barely five feet tall, with teeth and claws of vestigial vegetarians, and little more than sticks and stones in the way of armament.

Who, then, was this bipedal creature striding so straight and cocky among the megabeasts of the African veldt? After reexamining the cave massacres in detail, a fellow South African paleontologist named C. K. Brain came back suggesting a chest-deflating alter ego to Dart's killer ape. Brain agreed that many of the cave bones had indeed been contributed by predators, but by more conventional, four-legged carnivores. He added one other observation that turned Dart's killer-ape hypothesis on its head. He produced a fragment of *Australopithecus* skull, much like the Taung child's, bearing two arresting puncture holes. Borrowing a bit of Dart's own showmanship, Brain produced the jaw of a fossil leopard and neatly fitted the cat's two lower fangs into the holes in the little ape-child's skull. The murderous little meat hunter had suddenly become the meat.

Man the Scavenger

Over the decades, the pendulum of opinion on humanity's predatorial heritage swung between the extremes, between Dart's meat-eating monster and Brain's skulking scavenger. But from either perspective, the artist's reconstruction of the human beginnings took on the richly hued background of blood red. There had undoubtedly been protohumans walking the mean streets of prehistoric East Africa. They had left behind hip bones and craniums whose attachments and orientations could only subscribe to an upright, vertical mode of existence. In the ancient ash bed of a Tanzanian volcano, they had left footprints more than three million years old, bipedaling across the plain of Laetoli. One way or another, the protohuman was out there competing in the open among big and dangerous beasts, and winning its share of the contests.

That fact left a world of speculation as to how on earth the little australopithecine ape-men had pulled it off. That they had lived large on the treeless spaces suggested they had sustained themselves on something more than a simple gathering of nuts and fruits, tubers and roots. The richest concentrations of calories on the plains were those giant packages of meat wandering the grasslands in herds. As hard as it was to imagine the little near-humans tackling zebra, gazelle, and wildebeest, never mind elephants—all the while fending off the saber-tooths and leopards and roaming gangs of hyenas—it was harder to imagine getting by without them.

In 1968, wildlife biologist George Schaller and anthropologist Gordon Lowther decided to test their speculations. They set out into the predator-rich plains of East Africa's Serengeti in search of carrion and game like two australopithecine scavengers, on foot and unarmed.

For one eight-day experiment, the two set up camp on the banks of the Mbalageti River, whose waters in the dry season drew the great herds of the Serengeti. Their front yard offered views of strolling elephants and curious lions observing them from the far bank. "A group of hominids could never have had it better," wrote Schaller. "That night the lions visited us, roaring nearby and stumbling over the guy ropes as they circled the tent."

Over the next week, the neophyte meat-seekers set off each dawn along the Mbalageti, steering clear of thickets and ravines where lions and Cape buffalo were most likely to lay hidden. They followed circling vultures to the leftovers of lions and disease, they watched cheetahs bring down gazelles and plotted to steal their kills. They roamed as two red-blooded australopithecine scavengers in search of sustenance.

Schaller and Lowther settled into their primordial past, stalking with muscles taut, scanning the landscape for lurking danger and sizing up trees for emergency escape. They began to relish their daily adrenaline buzz, the air electrically charged with the hunt. In their heady thrill-seekers' euphoria, they grew careless. Seven times in the first two days, they found themselves beating hasty retreats after stumbling upon prides of dozing lions. "It is often said that one should never run from a lion," Schaller recounted. "This is nonsense." They began to carry protection, loading one of Lowther's empty beer cans with pebbles, and rattling their arrival whenever nearing potential lairs and blinds of ambush.

There were bad days burdened by hunger and the dejecting sight of bones picked clean. Then, there were days that "would have gladdened the heart of the most morose hominid." On one such day, Schaller and Lowther came upon an abandoned zebra foal. The little zebra was sick and slow afoot and stumbled to the ground after a short chase. Schaller stood over his prey and with a tug on its tail administered a symbolic coup de grace before letting the doomed foal return to its herd. The two later spied a giraffe calf, which oddly allowed Schaller to tiptoe within arm's length. Schaller looked up to find the giraffe staring ahead through a pair of opaque, sightless eyes.

Between bagging the sick zebra and the blind giraffe, the two slowest predators on the Serengeti had in one leisurely half-day's wandering procured what would have amounted to more than three hundred pounds of protein. The next day they came upon an old buffalo bull, dead of disease and truly ripe for the pickings. "Although hyenas and vultures had eaten the viscera and much of the rump, a great deal of meat remained, rather putrid but nevertheless edible," Schaller noted.

Schaller would later write, "There is neither cruelty nor compassion in a lion's quest for food and this impersonal endeavor strikes a responsive chord in man the hunter. I enjoyed watching most hunts as struggles of life and death at their most elemental. It is a time when each animal uses to the utmost those attributes with which evolution has endowed it. It is also a moment when man, weak of body and slow of foot, can watch his limits transgressed, a moment which engenders not only humility for his own lack of prowess but also pride and exultation that he has managed to survive at all."

The australopithecine experiment had given visceral support for the possibilities of early man as the lone predatorial biped of the African plains. With the scavenging lifestyle came ready explanations for human hallmarks such as family groups and divisions of labor (more hands to quickly tackle, butcher, and cache meat; more eyes to note incoming enemies). With the predatory component came an answer for the rudiments of speech and the process of planning. ("Look out, leopard in the bush!" or "Meet me on the far side of the water hole. Limping zebra there.")

Along with the ecological emergence of meat-chasing man-apes in the African plains came physical revolutions. The jutting jaw gradually flattened, the cheek teeth began to shrink. And the volume of the brain all but erupted. In the relative heartbeat of a million years, between the chimplike australopithecines and the human form of Homo sapiens, the size of the brain tripled. It had become the fastest-growing organ in the history of life.

That brain, along with the stone and bone tools it produced, would indeed come in handy for the hunting life. But it turns out the most important vehicles blazing the hunter's path may have been his own legs. When, in the 1980s, anthropologists began questioning the human stereotype as a hopeless slowpoke plodding about the plains, they uncovered one of nature's purest talents for pursuit. Hunting cultures were to be found running down the fleetest quadrupeds on the planet. A deconstruction of ancient human anatomy and physiology revealed one of the greatest long-distance runners of all time.

As the ancestral humans left the trees and hit the ground walking, their physiques and physiologies responded in kind. The bipeds grew tall and erect, their heads bobbing freely atop a long neck anchored to a stable platform of broadening shoulders. Leg bones lengthened, arms shortened, hips and waist narrowed. Midsoles of the feet began to arch; additional spring was added with the arrival of the Achilles tendon. They were the locomotory antithesis of the pig-necked, bowlegged, knuckle-shuffling chimpanzee life-form.

In a race with the furred and four-legged, the naked ape also ran cooler and more consistently than the competition. The running hominid vented heat not only through the panting mouth but also through the evaporative cooling from the sweatiest skin on the savanna. Running erect heightened the thermal advantage, exposing a minimum of bodily surface area to the sun.

These were just a few of the examined traits that padded the Homo sapiens racing pedigree. The contests were, admittedly, more of a tortoise-and-hare affair, with the slow and steady human ultimately gaining the advantage. But long-distance endurance may well have meant the difference between a life tied to the trees and a career as a big-brained biped competing in the carnivores' kingdom. Homo sapiens was born to run.

The marathon-man hypothesis came replete with anecdotes of modern cultures still known to run for their supper: North American Navajos chasing pronghorn antelope to exhaustion, Australian Aborigines running down kangaroos, African Bushmen overtaking zebra and wildebeest. Once looked upon with suspicion, the tales took a leap in veracity in the 1990s through the adventures of an anthropologist named Louis Liebenberg. Observing Bushmen of the Kalahari Desert of Africa running prey to death, Liebenberg went the distance, in a scientific exercise that nearly cost him his life.

The Bushmen's hunt would take place during the hottest times of the day, which in the Kalahari reached 108 degrees Fahrenheit. The runners, typically a team of three or four, would tank up on water, locate their quarry—a kudu, eland, gemsbok, or any of the various and fleet antelope of the Bushmen's world—and begin the chase. They would track their prey through long empty stretches of sand and confusing thickets, slowing only to regain a missing trail. Running for hours, they sometimes covered more than twenty miles. "When the hunter finally runs the animal to exhaustion," Liebenberg reported, "it loses its will to flee and either drops to the ground or just stands looking at the approaching hunter with glazed eyes."

Such exploits were not for the amateur. On one of the hunts, the observer Liebenberg abandoned his motor vehicle, went running with the Bushmen, and in the heat of the pursuit, forgot to check his own thermometer.

"By the time I caught up with !Nate, I was no longer sweating," wrote Liebenberg. "The hunters immediately recognized the early symptoms of heat stroke, and after having run down the kudu !Nate ran eight kilometers back to the camp to get his father, !Nam!kabe, to bring water . . . Afterward they explained that when running down an animal the hunter must continuously compare the condition of his own body with that of the animal, and I had become too focused on the animal."

Man the Hunted

In the early 1990s, at about the time Liebenberg was confirming for himself the human as fearsome pursuit predator, a Ph.D. primatologist at Washington University in St. Louis, had begun to ponder that same creature's alter ego as an item of prey. Donna Hart, a disciple of Brain's man-the-hunted camp, had come to the study of primates with a deep background in predator biology and, along with her adviser, the anthropologist Robert Sussman, a suspicion of one of primatology's fundamental assumptions: Predation, in the evolutionary scheme of primate life, didn't much matter. "Predation was just baldly discounted," said Hart. "The assumption was, 'No, it doesn't happen, because nobody's seen it.'"

The nonbelievers had apparently been looking in the wrong places at the wrong times. Hart dug up five hundred eyewitness accounts of monkeys and apes attacked by a range of predators, from raptors and reptiles to big cats, wolves, and bears. She added up the evidence of primate remains from the nests of raptors and the feces of carnivores. One leopard scat in particular— so compelling it was photographed—featured the undigested toe of a gorilla. As word of her search got around, unsolicited reports and testimonials from fellow researchers started piling in. In order to finally finish her dissertation, Hart was forced to close her files. Her tally of documented primate deaths by teeth and talon had ballooned to 3,600 cases.

Predation happened. And it didn't need to happen often to leave its mark. As John Terborgh had impishly noted in his 1983 monograph on societies of South American monkeys, "Successful predation is a rare event—at most it can occur only once in the lifetime of a prey." Suppose a troop of capuchin monkeys, Terborgh postulated, lost only one infant a year to, say, a harpy eagle. For a troop typically producing but one or two infants per year, such a loss would be anything but trivial. Such a loss couldn't help but change the way capuchin society conducted itself.

Hart's suspicions led her eventually to the most supposedly invulnerable of primate species. It turned out that one of the more compelling arguments for man the hunted had come ironically from the very skull on which the legend of man the hunter had been born. In 1995, the paleoanthropologists Lee R. Berger and Ron Clarke reopened the case of the famous Taung child, the two-million-year-old australopithecine toddler whose association with bashed-in baboon skulls had led Raymond Dart to label its kind "an animal-hunting, flesh-eating, shell-cracking, and bone-breaking ape." Dart had offered graphic scenarios of australopithecine hands prying away skull plates and reaching in for brain food.

When Berger and Clarke reexamined the Taung child's supposed victims, they came back with a scenario as gripping as it was contrary to Dart's. Berger and Clarke came away convinced the damages Dart had so assiduously ascribed to killer apes, and to the Taung child itself, were in fact the work of a killer eagle.

Berger and Clarke found modern passages of monkey predation that put the Taung child's hypothetical demise in a grisly new light. One report on a particularly ferocious African raptor called the black eagle, offered this:

When a complete animal is brought the eagle removes [the gut] first, which is not eaten, and then usually eats from the head. The eyes are eaten, and the tongue; the jaw is forced open so that the eagle can penetrate through the palate to the brain, which is also eaten. The scalp is removed from the cranium; the skin is removed from the face, chin and neck and eaten, together with the ears. The head is then broken off at the neck by twisting the neck vertebrae and eating the disks between them. Long strings of the spinal cord and bone marrow are pulled out and eaten, so are the thin shreds of the stomach and intestines. The heart, lungs and liver are eaten, together with the ribs, legs, feet and most of the vertebrae.

By way of the raptor's methodical butchery, the skull of the victim often bared telltale patterns of puncture and breakage. It struck Berger and Clarke that some of the baboon skulls from the Taung cave lacked the chew marks and paired tooth holes more typical of the leopard. Some had received small piercings in the braincase and cracking at the base of the cranium, which was more in line with talon damage.

"We would not presume to guess at the exact species of eagle responsible for the collection of the Taung assemblage," Berger and Clarke noted, adding what would become a prophetic conclusion, "although the crowned eagle is a likely candidate."

Stephanoaetus coronatus, the African crowned eagle, was a frightening, war-bonneted raptor with a scythe beak and fierce mien. Thick, cable-steel ankles powered heart-piercing talons. The crowned eagle featured on its trophy rack a broad range of animals as heavy as sixty-six pounds, including in East Africa a six-year-old child. Stephanoaetus coronatus was an aerial monster capable of regular and impressive carnage on organisms five times its weight.

The raptorial hypothesis was immediately met with one heavy objection. The Taung toddler, three to four years old and weighing maybe twenty-five pounds, would have been hefty cargo for even the most powerful eagle to lift off. Not so, countered Berger, noting that the eagle had likely disemboweled and partly dismembered the child as it would any other large primate, thus considerably lightening its load. It was not unknown for eagles to be seen carrying fifteen pounds of prey through the forest.

Convinced as he was of the Taung baby's killer, Lee Berger lacked enough hard evidence to convict. But then, ten years after airing his hypothesis, came a breakthrough. Berger was asked to review a manuscript submitted from a trio of anthropologists, led by Scott McGraw of Ohio State University, who had been searching the nests of crowned eagles in the Tai Forest of the Ivory Coast. More than half of the twelve hundred bones they collected had belonged to monkeys, a good many of them from big, fanged, and formidable species averaging more than twenty-five pounds—the Taung child's estimated weight.

But it was the victims' skulls, and the particular brand of wreckage inflicted on them, that raised Berger's pulse. He continued reading: "Punctures from talons were found inside the orbital cavities . . . Incidents of isolated damage, including V-shaped punctures and 'can-opener' perforations . . ."

Berger all but sprinted back to the museum where lay the Taung specimen. He peered into the ancient child's eye sockets. And in his giddiness,
Berger almost dropped the world's most precious hominid skull on the
floor.

Next to the tear duct of what had once been the right eye of the Taung
child, Berger discovered a tiny hole; in the left eye socket he discovered "a
ragged 'tear'" He had handled the skull hundreds of times, as had so many
before him. And nobody had noticed what he now noticed. What the Mc-
Graw manuscript was describing as the indisputable assaults of an African
crowned eagle on the skulls of large monkeys precisely matched the wounds
Berger now found in the little ancestral human skull in his hands. Two million years ago, a professional killer with the capacity of a crowned eagle had
swooped from the sky and taken the life of the Taung child.

"This is the end of an eighty-year-old murder mystery," Berger declared
in a press conference. "We have proved conclusively and beyond a reasonable doubt, which would be accepted in a court of law, that the African
crowned eagle was the killer."

The conviction answered much more than the immediate question of
who or what had done it. "It shows it was not only big cats but also these
creatures from the air—aerial bombardment if you will—that our ances-
tors had to be afraid of," said Berger. "These were the stressors and stresses
that grew and shaped the human mind and formed our behavior today."
Those who had ever startled to the harmless shadow of a passing airplane
could now look into the telling eyes of the Taung child and begin to
understand why.

Man the Haunted

A long legacy of stalking and being stalked had left lasting imprints on the
human physique and psyche. As late as the twenty-first century, ten thousand years after the advent of agriculture and the slow dying of the chase,
phantom hunters of the African plains could still be found in feats of the
modern human athlete. There were legs and lungs that could cover twenty-
six miles in little over two hours. There were arms that could accurately hurl

a spear (disguised as an Olympic javelin) one hundred yards on the fly or throw a rock (fashioned into a five-ounce baseball) one hundred miles per hour.

In turn, those creatures that had hunted the hunting ape remained lurking, in the dark reptilian recesses of the subconscious mind. "Why do we jump when we see a large object that comes into our peripheral vision?" asked Donna Hart. "I don't think it's just because large blobs are what we fear. I think we have millions of years of history of seeing something large and dark in our peripheral vision that could very well eat us."

"Not getting killed is a powerful force to deal with each day," concluded Hart and Robert Sussman from their exhaustive foray into the archives of human predation. Yet for better or worse, the force was no longer. Except for a smattering of dangerous outposts, where human incursions were pressing against the carnivores' last wild bastions, death by predator was rarely an issue of daily concern. Tigers still stalked woodcutters in the Sundarban swamps of Bangladesh; wolves still carried away stray children from the village edge of rural India; crocodiles still snatched untold thousands of victims from the rivers and lakes of Africa and Australia. But beyond these and a few other anachronistic hot spots, man-eating had been relegated to the realm of nightmares and media circuses. When one day in 2004 a cougar in Southern California killed a mountain biker and badly mauled another—one of less than twenty lethal cougar encounters recorded in the country over the previous century—the news went national, the stories running for months thereafter. The indisputable tragedy of those maulings notwithstanding, many more Americans died that year by dog bite. Yet it was the lone wild predator that got the monster's treatment.

Those who'd spent their lives pondering the impacts of the topmost predators could only speculate on where the human species was heading without them. "We don't have to constantly look over our shoulder, we don't have to live behind lion-proof fences, our kids can play in the garden without mothers being worried," ventured Hans Kruuk, the esteemed animal behaviorist and confessed carnivore addict. "Our observation and detection abilities do not need to be as keen and alert anymore, nor do we need to be able to sprint—and with all such faculties, it is known that if you don't use it, you lose it. One can afford to be an obese couch-potato."

Others were not quite so comforting with their prognoses.

"If the big carnivorous creatures did not exist, the world would be an emotionally and esthetically impoverished world," said Stephen Kellert, a social ecologist from Yale. "Stripped of their menace, powerful creatures and landscapes become little more than objects of amusement and conde-scension."

George Schaller, in summing up his Serengeti experiences, wrote, "Now, in our unheeding rush to conquer our environment, we are in danger of destroying the roots of our nature, the wilderness that saw the whole of our evolutionary history. Perhaps as we transcend our past, adapting to new pat-terns of culture and becoming less human by today's standards, the wilder-ness will become superfluous. But for the present the salvation of our humanity lies in the spirit of such areas as the Serengeti, where man can re-new his ancient ties and ponder his uncertain destiny."

"In view of the enormous scope of human time and experience, perhaps mankind has unwittingly embraced a diseased era as the model of human life," wrote Paul Shepard. Before his death in 1996, the eco-philosopher Shepard had come to view modern man as a lost and lonely refugee from the Pleistocene. Estranged from his predatorial heritage, indifferent to the creatures that had made him human, the adolescent heir of the natural world was plundering and blundering his way to the gravest mistake of his-tory. "The most damaging blows of all are the extinctions of the 'useless' forms of life, those wild things that seem outside our economy and inimi-cal to agriculture . . . We have loosed a population epidemic since men ceased to hunt and gather that is the most terrifying phenomenon of the million years of human experience."

Prairie Ghosts and Phantom Pains

Though he died ten years before Pleistocene rewilding made international headlines, Paul Shepard loomed large in its philosophies. "Perhaps nothing makes us more human, and feel more vulnerable, than a large predator that could kill and eat us," concluded Donlan and Greene in summing up the mob scene they'd so innocently incited.

Several of the rewilding authors later mentioned in private that they might have erred. Not with any scientific detail or train of logic, but with their hesitation to more proudly trumpet the gut-level grandeur of their vision. Buried beneath the more mechanical discussions of evolutionary potentials and economic costs and gains, lay hints of heart-stirring drama and visceral inspiration. "Free-roaming, managed cheetahs in the southwestern United States," began one formal passage appearing in their incendiary *Nature* paper, "could save the fastest carnivore from extinction, restore what must have been strong interactions with pronghorn, and facilitate ecotourism as an economic alternative for ranchers." Following is the story left untold.

It turned out that one of the most astounding creations of the Pleistocene was still around to ponder. The American pronghorn, a strange and singular mammal of the American plains and badlands, is more closely related to the goat than to the antelope for which it is most commonly misnamed. The pronghorn is a barrel-bodied creature on pencil-thin legs, with a thick neck and long face and stout, namesake horns sprouting from the crowns of mature bucks. Viewed broadside, it is colored by a swath of carmeled prairie, followed by a flaring white rump. Back from its near extermination during the shooting-gallery era of the American West, the pronghorn again epitomizes the open expanses of the American prairie. Modern biologists like to pry into its food habits and population dynamics and harem mating society, but almost everything written or said about the pronghorn invariably circles back to one singular overriding theme: speed.

Capable of traveling at bursts exceeding sixty miles per hour, the pronghorn is the second-fastest land animal on the planet—or the fastest, depending how far the race is run. Its sprint is slightly slower than that of the African cheetah's, but its pace over the mile is unmatched by any wild creature on legs. Its feats of footspeed are both legendary and true. Many who have spent any appreciable time driving through pronghorn country come back with a nuanced version of the same ensuing spectacle.

It begins in the driver's seat of a pickup truck on a dusty road, far upon some open stretch of the American steppe, when out of the corner of the eye appears a band of pronghorn on the run—stick-legs a blur, white rumps shining, muscled necks craning forward. Over the sagebrush badlands they fly, a squadron of hovercraft fluidly absorbing the terrain, hurtling forward at a frightening clip. The speedometer confirms what the eye struggles to fathom: forty-five miles per hour. The pronghorn are keeping pace, unveering, unflagging. One minute, two minutes—they are still going. They are racing the machine.

Several anomalies of anatomy and physiology account for the pronghorn's supernatural speed. The lower legs are long but only slightly thicker than a human index finger, surprisingly slender even for an animal barely exceeding one hundred pounds. Those legs gobble twenty-nine feet of ground with every springing stride of a pronghorn at speed. Probing inside the pronghorn body, the physiologist gasps: Heart, lungs, and windpipe are freakishly oversized. A tame pair of young pronghorns are placed on a treadmill, oxygen masks strapped to their faces, and tested for the volumes processed by those tremendous organs. They triple the capacity of a comparably sized goat.

Behind such outrageous displays of velocity are fearsome invisible forces. To see them, one must think back at least thirteen thousand years ago, before their disappearance. Before then the plains of North America were stampeded by an unprecedented cast of quick and deadly predators: a lion larger than its contemporary African subspecies; wolves of several varieties; a bear designed like a racehorse; a hyena with the legs of a coursing hound; and a cheetah with the legs of, well, a cheetah. Miracinonyx trumani was lithe yet larger than the modern African cheetah, and likely at least as fast. It was this sprinting cat, and the formidable packs of hyenas and the like, that made the American steppe a very lively place to grow up. They were the crucible in which the ultimate demon of speed, the American pronghorn, was forged.

That at least is the theory of pronghorn expert John Byers, along with many others who can think of little reason to refute it. There is nothing in the modern American bestiary that comes close to challenging a running pronghorn that's beyond a few weeks old. It is, in Byers view, overbuilt.

The pronghorn has apparently not lost more than a step or two in the dozen millennia since the last of its most capable chasers departed. Ironically enough, one of the many names given to the pronghorn is prairie ghost. Earned for the speedster's habit of vanishing into the open spaces, it might just as well be referring to the invisible cheetah on its heels. Through a fortuitous oversight of evolution, those with imagination can still peer back into the Pleistocene, through the blazing speed of the pronghorn, to where magnificent carnivores still wielded their power, crafting some of the most awesome inventions in all of nature.

The pronghorn's was one of many such stories finding few sympathies among the rewilders' audience. Naysayers at their polite best chided the rewilders for romanticizing the past; at their sniping worst, for tempting a *Jurassic Park* disaster. To these the rewilders quietly voiced a sad and stinging reply. The most dangerous experiment is already underway. The future most to be feared is the one now dictated by the status quo. In vanquishing our most fearsome beasts from the modern world, we have released worse monsters from the compound. They come in disarmingly meek and insidious forms, in chewing plagues of hoofed beasts and sweeping hordes of rats and cats and second-order predators. They come in the form of denuded seascapes and barren forests, ruled by jellyfish and urchins, killer deer and sociopathic monkeys. They come as haunting demons of the human mind. In conquering the fearsome beasts, the conquerors had unwittingly orphaned themselves.

There comes a gradual bend in the road, sending the veering truck and beelining pronghorn toward a collision. The pronghorn decide they will be first to the intersection. The whirring legs accelerate. With a boost from some internal rocket, they surge ahead. They unzip the road in a puff of dust and silent clap of thunder, heading for far horizons, white rumps narrowing to white dots in the distance.

In the hurtling pronghorn, the vanished predators have left behind a heartrending spectacle. Through the smoking displays of wild abandon runs a desperate spirit, resigned to racing pickup trucks in its eternal longing for cheetahs.

EPILOGUE   Alone on the Hill

NEARLY FIFTY YEARS AGO, when the trio of Hairston, Smith, and Slobodkin first raised the hackles—suggesting that the world was green by the predators' good graces—there was much to be argued. There still is. Great keystone predators do not lurk benevolently behind every rock, nor does their presence cure every ecological ill—at least not that science yet knows.

However near or far HSS came to generating a universal theory for how the web of life is woven, they nonetheless stimulated a world of inquiry. Nearly fifty years later, the community of ecologists knows that there are indeed certain keystone predators, some as unlikely as an orange starfish, with powers to rock entire ecosystems to their foundations. It knows that the fertile kelp forests of the North Pacific coast are far more lush and lively when sea otters are around to eat urchins—though it is fiercely divided as to whether maybe, just maybe, a gluttonous legacy of industrial whaling ultimately triggered a bizarre cascade of killing that once again has the otters and their forests crashing. The field of ecology, if it is not yet entirely sure, is growing more convinced with every new skyward aspen or willow in Yellowstone that the greening of the nation's most hallowed park has much to do with the return of its missing wolves. The field has also begun to take seriously the prospect—though not as seriously or urgently as John Terborgh would like—that the ecological meltdown visited upon the predator-free islands of Lago Guri may well soon describe broader hells on Earth.

"Posterity has rightly treated HSS kindly," declared the benevolent godfather of keystone ecology, Robert T. Paine, in a retrospective essay on the fortieth anniversary of the paper that sent the field afield. It seemed almost anywhere one chose to look, arctic tundra to tropical forest, one could find ecosystems ascending to higher orders of biological stability and complexity, or teetering toward chaos and impoverishment, in varying accord with their apex predators.

By the fall of 2007, Terborgh was organizing an international guest list, a who's who in this flourishing field of ecological cascades, to gather for a February conference on what he had declared as "the next hot topic in conservation." The exploration that had begun with the question Do the great predators matter? had come to the more complicated task of dealing with the answer.

In so far as this has been a story about the science of vanishing predators, here is where the story should logically end. Conserving big predators, and its parent discipline, eradicating them, is where the science of top-predator ecology gives way to myth and sentiment, heroics and hypocrisies. Whichever side one stands on bringing back dangerous, meat-eating beasts, there is a safe bet that logic and systematic reasoning have little to do with their ultimate position. "It is usually us," said the behaviorist Hans Kruuk, "mankind, who decide what kind of ecosystem we want to have in the first place, in an esthetic—not scientific—decision. So whether we want to have large carnivores around because they appeal to us, or because they play an important role in an ecosystem that appeals to us, in the end this is an esthetic decision. And in this, I believe that the appeal of the animals themselves will be the strongest."

It does on first glance appear the predators' fates are lately brightening with fresh displays of our tolerance. In North America a spate of top carnivores are being helped, or at least allowed, to return to many places they were once forcibly helped to leave. From the few dozen wolves flown down from Canada a decade ago, there are now more than a thousand roaming the northern Rockies of Wyoming and Idaho. After forty years of absence, lynx are again breeding in the San Juan range of Colorado, after being shipped all the way from the Yukon. In the Blue Range straddling Arizona and New Mexico, there are now about sixty Mexican wolves, descendants of five captive animals once representing the last Mexican wolves on Earth. Cougars are filling old voids in the American West and wandering eastward, with tantalizing rumors reaching all the way to the north woods of New England. Jaguars from Mexico are showing up in the desert highlands of the American Southwest. In New Jersey, the most densely peopled state in the United States, its citizens are meeting black bears with unheard-of frequency. And though there is good reason to believe that a lack of wolves is partly behind the success of the coyote, the scrappy song dog of the West has nonetheless become a cosmopolitan citizen of the nation, chasing geese across Chicago golf greens, denning within howling distance of the White House, leading cops on chases through New York's Central Park.

The predators of the sky also seem to be finding safer airspace of late. Since the shooting was stopped and egg-thinning pesticides were banned, the all-but-obliterated bald eagle is again nesting above many a river bend and bay shore. Peregrine falcons have been returned by hand to cliffs long abandoned—and beyond, now stooping from skyscraper eyries upon fat city pigeons. And as many a backyard bird-watcher can attest, with varying parts fascination and horror, the Cooper's hawk is back, with fierce red eyes ablaze, terrorizing suburban feeders and lawn flocks of robins, doves, starlings, and cowbirds.

Such revivals are not solely New World phenomena. There's a European renaissance underway in the Alps, where the rural farming populace is vacating the countryside for metropolitan employment, and the long-banished bears and wolves and lynx are tiptoeing back.

And now, in all fairness, another side of modern predator aesthetics, from a more dominant majority: Off the coast of North Carolina, to pick just one recent reminder, the bull sharks and tiger sharks, blacktips and hammerheads have over the past thirty years been fished to but a single percent or three of their former abundance. And in the great sharks' absence, their prey has multiplied by orders of magnitude. The waters have grown thick with smaller sharks and their cousins, the skates and rays. One in particular, the cownose ray, prefers to eat shellfish. Behind the massive new schools of some forty million cownose rays now sweeping the eastern seaboard, there comes word of declining commercial hauls of clams and oysters. With the big sharks' annihilation, and the rays' release, North Carolina's century-old bay scallop fishery has collapsed.

And so, taking heed of these ominous warnings from science, fishermen the world over have now agreed to end their global persecution of sharks.

Well, no, actually nothing of the sort. In the piratical domain of the open seas, up to seventy-three million large sharks each year are being hauled up and hacked alive for little more than their fins, to feed wealthy Asians' appetite for shark-fin soup. Esthetically speaking, few seem particularly mindful, let alone concerned, of the modern ocean's conspicuous void of big, flesh-eating fish. These are waters fast giving way to urchins and jellyfish, algae and bacteria—the bottom-most bottoms of the food chain, the microbial masses—what marine biologist Jeremy Jackson prefers to call "the rise of slime." In Jackson's unvarnished observations, jellyfish have become the catch of the day, and slime the future of the oceans.

Not only does the slime rise, but over time it actually begins to look *good*. Slime has become the norm in many young minds (among them young conservation biologists). And the younger the observer, the more acceptable the slime. It is the phenomenon made known by the marine biologist Daniel Pauly as the shifting-baseline syndrome. The world as first seen by the child becomes his lifelong standard of excellence, mindless of the fact he is admiring the ruins of his parents. Generation to generation, the natural world decays, the ratchet of perception tightens. Gradually, imperceptibly, big sharks give way to small sharks, small sharks to baitfish, baitfish to jellyfish to slime. On land, the big cats and wolves become feral house cats and coyotes. The wild standard sinks ever lower and becomes ever heavier to raise. Few notice, few care. Eventually, nobody remembers that wolves not long ago freely roamed the Adirondacks, and hence there is mad howling over the suggestion of returning them to their homeland. Southern Californians panic on learning that a cougar track has been discovered on the fringes of their gated neighborhood—mindless that cougars roamed these hills and canyons long before gated communities drew their lines in the chaparral. Shifting baselines help explain why the Pennsylvania deer hunter sees everything right and nothing wrong in a forest that's swarming with deer yet as barren of biodiversity as a city park. Slime creeps in many guises.

It is a maxim among carnivore biologists that the main reason big predators now die is because people kill them. People run them over with speeding cars, they shoot them, trap them, gas them, poison them, and torture them. Laws notwithstanding, they kill them out of sheer spite, then bury the evidence—a practice so routine it has become a barstool commandment in certain rural cultures: Shoot, shovel, and shut up.

The corollary to the man-as-death maxim suggests humanity also serving as last-minute savior. It is a comforting salve for the guilty conscience. We got them in trouble, we can get them out. More than twenty years ago the eminent biologist Edward O. Wilson identified an inherent trait in all of us, an "innate emotional attachment to other living things—a love of life—biophilia." Biophilia, unleashed, could ostensibly come to the rescue of those forsaken beasts now cowering in the far corners. But with one big catch: Biophilia, it seems, requires care and feeding. One must engage the world of living things, the sooner in life the better, for the dormant seeds of the biophiliac to germinate.

And if there is a biophiliac revolution in the making, it does not appear the next generation is training well for the task. The average American child today spends all of thirty minutes each week playing and exploring outdoors. The same child spends more than ten times that amount wandering the wilderness of the World Wide Web. Add to that the whole suite of electronic media—the home and theater movies, the video games—and the screening hours nearly double. Patricia Zaradic and Oliver Pergams, two particularly concerned conservation biologists, have a name for this emerging "human tendency to focus on sedentary activities involving electronic media." They call it videophilia. Zaradic and Pergams have been wondering just what videophilia offers the future of nature. If the time spent in family outings is any indication, videophilia is a stultifying, eye-glazing disaster. Since 1987, after decades of steadily increasing popularity, U.S. national park attendance has withered 25 percent—a drop uncannily mirroring the rise of the Internet and video revolution.

"The greatest threat to conservation and to the environmental legacy represented by the U.S. national park system may be more subtle than bull-dozers and chainsaws," the two write. "If we are indeed seeing a funda-mental decline in people's appreciation of (and attachment to) natural areas, the authors feel this does not bode well for the future of biodiversity conservation."

There are others who would like to believe whatever role the bygone car-nivores may have played in the rise of creations over the eons is now being nobly filled by their omnipotent, last-minute substitute, the human hunter. That, at any rate, is an argument made familiar by a certain class of sports-men and their game agencies. Leave the killing to us, goes the mantra, and we will balance the herds. Given the possibilities that the rifle-armed hunter may one day soon remain the last predator in the woods, it is a postulate worth examining: Are humans now functionally equivalent to large mam-malian carnivores?

We certainly know our weaponry is of a different plane of killing power. We now have hunting bows propelling arrows at three hundred feet per second. Modern hunting rifles fire bullets at three times the speed of sound. With those kinds of ballistics and a high-powered scope, a rifleman needs stalk no closer than a quarter mile to bag a 750-pound elk that never knew what hit it. For that matter, a hunter armed with pitfall traps and cable snares can kill elephants in his sleep.

Examinations of big-game hunters in Africa and North America come to similar and unastounding conclusions. Sport hunters tend to go heavily for trophies, selecting the biggest, handsomest, fittest bulls and bucks—skimming the cream of the genetic crop. Working carnivores, on the other paw, tend to take the young, the old, the lame, and the weak, with efficiency of effort and immediate survival foremost in mind. Sport hunters tend to concentrate their kill in a few weeks of the regulated season, which means an elk in wolfless Col-orado has ten or eleven months between hunting seasons to make a clear-cut or wallow of whatever streamside grove they care to lounge in. It means scavengers like grizzlies—if there were any left in Colorado—would be more hard-pressed to feed their young in spring if the only carcasses were to be found in the wake of rifle hunters in the fall.

Wolves, it has also been found, do not build roads into wilderness areas, chase animals to exhaustion on ATVs, howl at ninety decibels for hours on end, compact and erode soils, cave in stream banks, tear up meadows, crush plants, or foul the air and water with hydrocarbons.

With pardons for whatever sarcasm may appear to have seeped into this discussion, these are in fact distinctions listed in serious, straight-faced studies, including one by mammalogist Joel Berger. The goal of managed hunting, concludes Berger, remains founded on human profit, not the preservation and stability of biological diversity. And Berger is not holding out great hopes for some miraculous resurgence of biophilia to save the day. "Perhaps the best we can do is recognize differences imposed by our own human culture and our hunting, and attempt to maintain places that are good for our souls."

Those most soulful places are going fast. Berger's home range in the heavenly regarded high country of the Greater Yellowstone Ecosystem is one of the handier examples. Sprawl—and its attendant roads, fences, four-wheelers, pets, diseases, weeds, and other associates of ecological decay—has taken huge bites out of the predators' Greater Yellowstone sanctuary. From 1970 to 1999, the ecosystem's human population bulged by nearly 60 percent. Wealthy young professionals and retirees with their luxury ranchettes swarmed upon Yellowstone's riversides, hillsides, and mountain-tops. The acreage of rural land under development nearly quadrupled.

Yellowstone, with respect to sprawl, is hardly exceptional. The prime habitat of the New West has become a château with million-dollar panoramas bordering a national park. At a time when large carnivores might be looking to regain lost ground, the former wildlands of the West are rapidly becoming far less friendly neighborhoods.

Cynics argue that Yellowstone has become a zoo, and the large carnivores its captives. If one considers the cultural climate for predators just outside the gates, the cynics have a strong argument. A roaming wolf pack runs afoul of a cattle rancher, and soon there is a government gunner in the air taking every last pack member out. A grizzly bear gets accustomed to finding free meat following the report of an elk hunter's rifle, and eventually the bear gets a bullet of its own.

These days there are planes and helicopters aloft over North America, gunning for the predators by the tens of thousands. There is also a corps of dedicated agents on the ground, legally sanctioned and otherwise, setting traps and broadcasting poisons. Their sponsors are largely U.S. taxpayers (many of them unwitting), their strategists of the opinion they are sparing hoofed game for sportsmen, hoofed livestock for open-range ranchers, and peace of mind for humanity. Though researchers have for decades been panning such scattershot practices as monumental wastes of life and money—recommending instead nonlethal alternatives and more selective targeting of true culprits—the rangelands remain mined and the skies continue raining bullets. In the particular case of the coyote, the most conservative body counts consistently run to more than seventy thousand per year.

Among those of the antipredator persuasion, neither rarity nor science nor popular vote count for much in deciding the predators' fates. In 2007, twelve years after gray wolves were reintroduced with majority endorsement to the backcountry of central Idaho, the governor of the state, Butch Otter, publicly declared his desire to see all but the minimum number of one hundred wolves killed, adding he'd like "to bid for that first ticket to shoot a wolf myself." Wyoming, for its part, has proposed reclassifying its wolves as predators, so its citizens could shoot them on sight. In Oregon, commissioners from Jackson County hired a houndsman to spend half his year killing mountain lions—supposedly for the safety of the people in a state that had no record of attacks nor sound evidence to suspect any forthcoming.

Not to single out the United States—the loathing encircles the globe. Killing predators, common or rare, is still practiced wherever the predators can still be found. Cheetahs in Namibia, snow leopards in India, wild dogs in South Africa, jaguars in Mexico, pumas in Patagonia, are all under the gun. In Kenya, young Masai warriors, who once proved their manhood by killing lions with a spear, these days more often leave the job to poison. Remember the good news about wolves and lynx and bears coming home to their historic haunts in the Alps? Several years ago in Europe, during part of that same repatriation, a celebrated brown bear crossed from the Italian Alps to become the first wild bruin in Germany since 1835. Bruno, as he was named, became a popular but mischievous bear, with an inclination to lounge in public places and an unfortunate taste for sheep. When the authorities grew weary of Bruno dodging their attempts to humanely trap him, they shot him dead.

There is a tendency to read in the comeback headlines a sense of mission accomplished. The first wolves here in so many years, the first bears there, mountain lions in the backyard—primal nature on the mend! And that is one of the dangerous fallacies the discoveries of carnivore ecology has exposed. It is no longer enough to parade a few token predators in a landscape that needs hundreds to work their magic. A single pack of well-placed wolves in Yellowstone could probably thrill the same gathering crowds on the Lamar Valley hillsides, but those few wolves would be far beneath the task of policing every corner of the park against the inevitable lootings of too many elk. When bureaucrats boast about success of the Endangered Species Act, and therewith yank the Yellowstone wolf and grizzly off the list, they are merely reopening the door for the slaughters of old, as the recent dispatches from Idaho and Wyoming confirm. More vitally, they are expressing a blind ignorance on the emerging science of the predator's irreplaceable ecological place. Even a park full of wolves, huddled in the protected confines of Yellowstone, leaves the other nine tenths of the Greater Yellowstone Ecosystem untended from the top down. It leaves a bleaker semblance of the wildness it is falsely lauded to be. Says Michael Soulé, who wrote the book on the crisis discipline of conservation biology, "I strongly believe unless we do get our large carnivores back into our ecosystems, that they are going to continue to be degraded over time, no matter what we do with connectivity and protected areas."

Before believing either the cheering applause or the doomsday protests over all the wolves and grizzlies supposedly storming the streets of America, it is instructive to mull a less sexy fact or two: Neither species has yet regained 5 percent of its historical range in the United States, which is, for all but a few pushpins on the map, a nation increasingly overrun by people and prey species.

Enough already. This was never intended to become an elegy. This was to have been a celebration of discovery, an awakening to the long forgotten and yet unknown powers of the great predators. And so, beyond the grayness that inevitably pervades a story of missing limbs and phantom pains, here are yet a few patches of blue sky worth admiring while they last.

At one of the largest sheep ranges in Idaho, more than eight thousand of the planet's most helpless and inviting prey animals are grazing amid a land thick with dangerous predators, and living to tell about it (so to speak). After a deadly run-in one night with a pack of wolves that took twenty-five of his sheep, Mike Stevens, president of Lava Lake Land and Livestock, now has his sheep sleeping safely in portable corrals ringed by solar-powered electric fences, and further fortified by massive Great Pyrenees sheep dogs. Stevens's human shepherds carry receivers that tell them when any of Idaho's radio-collared wolves may be roaming the area. "I basically told our foreman that I would fire anybody that shot a wolf," said Stevens. When wolves are caught snooping, they're sent running with shotgun cracker shells and rubber bullets. They are taught a jarring but bloodless lesson that sheep are disagreeable game.

"Our vision is that we are co-owners and operators in one of America's great wilderness areas," said Stevens. Lava Lake's business is lamb and wool and landscape-scale conservation, and so far, it seems to be working—wolves, sheep, coyotes, shepherds, and all.

Some of the money for Lava Lake's high-tech, deep-ecology approach to ranching comes from the nonprofit organization Defenders of Wildlife. The Defenders, for their part, back their pro-carnivore agenda with hard cash, by compensating those who work in predator country. Since the wolves began roaming the Rockies of Wyoming and Idaho, the Defenders have paid out a million dollars to help ranchers keep their livestock safe without killing predators (as in Lava Lake's case) or recoup the losses that sporadically crop up.

Here is another feel-good story from a little town on the prairie, east of Montana's Rocky Mountain Front. The town is Choteau, an extraordinary place in space and time, because this is where the grizzlies, which once roamed the plains when Lewis and Clark came through two centuries ago, are returning.

The grizzly is coming home. And here is the kicker: This time nobody is shooting them. Here is a place where the bears are being treated with the same tough love as the Lava Lake wolves, with rubber bullets and barking dogs and the message that they may pass but not linger. Here's a place where cattlemen are having their old dead cows hauled off to the back-country, so the bears don't get used to looking for food and finding trouble in their corrals. Here's a place where they don't send out for the National Guard when a young bear happens to wander into town while grade school is in session; they postpone recess until the bears are escorted away.

This is still very much a rural community that forty years ago would have shot those grizzlies. In Choteau, people and bears are coming to a mutual understanding that, with a respectful distance, both can go about their lives in peace.

The great predators are great barometers of our maturity as a species. If we can live with an animal that could just as soon eat us as an apple, if we can make room for an animal that traverses entire states looking for a mate, how better to define the art of compassion?

There are yet more courageous displays of coexistence found in the few places where big dangerous carnivores and people still play the deadly game of Africa's ancient Eden. In Tanzania there are farmers who, while sleeping in fields to guard their crops from destruction by bushpigs, are sometimes dragged out of bed and eaten by lions. Yet even in this land where the worst nightmares of humankind sometimes come true, there are those still thankful for the lions for helping keep the bushpigs from their fields.

In tiger-menaced mangroves of Bangladesh, fishermen and woodcutters have taken to wearing a mask on the backs of their head to thwart the tigers' instinctive preference for surprise attacks. Instead of sending in the hit-squads, they have constructed mannequins armed with an electric charge that jolts the tigers into thinking twice before attacking the human life-form again.

It might be fair to mention here that some among the conservation community would begrudge the attention given the charismatic megafauna, for the precious few attentions they might steal away from the wee and incon-spicuous of the natural world. It is a criticism that's well intended but ulti-mately off target, in that nearly all wild species are given short shrift in the ever humanized landscape. The larger dilemma remains: Nearly none of

Earth's troubled plants or animals are getting what they need. In the 2007 iteration of the World Conservation Union's red list of threatened species, covering blue whales to wolves to woolly-stalked begonias, only 1 out of 16,306 species listed made significant progress, however dubiously: The Mauritius echo parakeet was downlisted from critically endangered to merely endangered. (The woolly-stalked begonia, for those curious, was declared extinct.)

There are also some who might have hoped for equal time in these pages for the little things that also run the world—and with all sympathies here extended. Creatures that one could pinch between two fingers collectively wield their own inordinate powers. But more to the point—let's bring it back to us—those little things account for so many of the everyday pleasantries of life as many of us know it. Butterflies, bees, bats, and the vast array of fellow pollinators are the artisans and laborers behind one of every three mouthfuls of food humanity eats. And that's not including the microfaunal service of decomposing and recycling the world's wastes, and their vicious warfare upon our crop pests. A world removed of smallness would soon have the human race nose-deep in waste, eating dirt in desperation. "If invertebrates were to disappear," declared E. O. Wilson—and the little things have never had a greater champion—"I doubt that the human species could last more than a few months."

There has also been no mention so far about the changing climate—that oncoming Armageddon of greenhouse gas that scientists have been warning of (and oil and automobile executives and their PR professionals covering up) for more than twenty years. The thickening atmosphere produced by the combustion-based lifestyle promises all sorts of surprises for every form of life, predator and prey. Unprecedented heat waves, bottomless drought, super storms, melting glaciers, biblical floods—in such forecasts of chaos, all bets are off on sparing the few lonely predators stranded at the top of the endangered heap. One might imagine a warmer world actually benefiting certain wolves, pressed as most of them now are into the most inhospitable ends of the frozen north. The grizzly bears of Yellowstone probably won't be so fortunate. If the warming were to bring plague to the whitebark pine, or otherwise push the tree out of the park, there would also go the whitebark pine nut, one of the grizzly's essential foods. The polar bear is another bad conservation bet in a greenhouse future. With its sea ice domain melting ever sooner and farther by the season, there are predictions that the polar bear's home may be nearly gone within fifty years.

And once again my apologies for that depressing dirge so relentlessly creeping back into the conversation.

There is, however, at least one more likely scenario to report, and as to whether it qualifies as good or grim, let the reader judge. It is an alternative scenario in which the carnivores do have a future, albeit in the form of a mechanically managed affair. To illustrate again with the workhorse model of Yellowstone, wolves and bears leaving the borders are getting killed with withering frequency in the human-dominated landscape. The animals' corridors of safe travel between refuge are being pinched, their options against inbreeding growing slim. And so to keep their blood lines strong, and their numbers viable for long-term survival, may one day soon become a matter of . . . trucking. This is a foreseeable future of big predators in pockets here and there, managed and bred more like livestock, trucked and airlifted between fenced reserves. It is what some are already considering as the only hope left.

"Conservation of large expanses of pristine habitat may be the utopia of carnivore conservationists, but sadly this is a possibility already lost to history," notes the carnivore biologist David MacDonald. "Nature as a theme park is an unattractive prospect, but there are parts of the world where the option of laissez-faire conservation is long past."

As any number of pragmatic biologists agree, what has not already been decided for the future of big carnivores will very soon be decided by us. Do we want them back? If not, we can say for sure what their future will be. As for the trickle-down consequences to come, that is something only time will tell. Life will probably not come crashing to a halt for the lack of big meat-eating beasts, at least not in ways that we're yet able to clearly define. "Is it an ecological impact we can survive?" ponders the conservation biologist David Wilcove. "Well, that's kind of a high bar, isn't it? If we only worry about the things that threaten our very survival we end up in a pretty bleak world. Of course, that's true not only for biological diversity, it's also true for culture, literature, and the arts."

For all the lip service and heartfelt longings to see the fierce beasts return, there often remains unspoken the reflexive shudder of relief that they are securely caged, buried, or banished to the hinterlands. Alfred Russel Wallace, the indefatigable nineteenth-century naturalist who nearly beat Darwin to the explanation of evolution, is often quoted for one of his prophetic exclamations on the great Pleistocene extinction. "We live in a zoologically impoverished world," wrote Wallace, "from which all the hugest, and fiercest, and strangest forms have recently disappeared." The quoter usually ends there (as did this one, much earlier in the story), leaving one to imagine Wallace wistfully longing for that wild world of yesterday. Yet the remainder of Wallace's thought, so often buried by historians, suggests another sentiment: ". . . and it is, no doubt, a much better world for us now they have gone."

"I suspect most people in the world could not care less if all large predators vanish," said George Schaller. "There are a few of us who think they are beautiful, interesting, essential to natural ecological processes, and part of our natural heritage worth preserving, but we are a distinct minority anywhere."

Among the many rewards for surviving the long, thrilling trial of the Pleistocene came the imparted wisdom to never take lightly those big, carnivorous creatures that with one swipe of the paw can kill and with the other instill a sense of grace and majesty to the art of survival. It is the durability of that lesson that will decide whether the dangerous beasts accompany the great human experiment into the twenty-second century. Writes Stephen Kellert, "The survival of tigers, bears, mighty rivers, and wilderness will depend as much on our avoiding and fearing these species and landscapes as on our willingness to shower them with affection." Some have muddled the lessons over time and distance. And so there are those who jump into lion cages or purposely pitch their tent in the middle of grizzly bear trails, just as there are others who shoot tame tigers for sport or saw the jaws off living coyotes. Somewhere in between lies hope, for whomever might be looking for it.

This story began with an admission of bias, and finishes there for good measure, in one rare and particular place that still harbors grizzly bears. The morning of June 21, 2000—summer solstice, longest day of the year—found me first thing on the trail heading back to the aspen grove on the hill, where the night before the grizzly family had burned itself into one person's psyche. That morning I discovered, exaggerations aside, that the world had changed.

It was a hike beginning on high alert, half expecting a grizzly behind every sagebrush, a bear rearing from every draw. But soon the legs loosened and the jitters gave way to a deepening state of fascination. There was a strange procession of tiny worms on the trail, one every foot or two, each wriggling in the dust, as if being attacked by an invisible enemy. In a green meadow on the flank of a distant butte walked a moose, disappearing slowly like an apparition into the trees. From a trailside sagebrush, a Brewer's blackbird scolded, impressively so without dropping the insect in its beak. Snipe winnowed from the sedges of the trickling creek, ground squirrels peeped and trilled as they raced across the path.

After a blissful hour on the trail came the final, watchful approach to the meadow where the family of bears had appeared the night before. There beside the aspen grove lay the broken ground, freshly gouged and pocked by the mother's foragings. Between deep green patches of geraniums and lupines lay chocolate scoops of dark earth. Impressions in the turned soil—heel, palm, toe, and claw—brought images of great paws deftly excavating earth by the bucketful, the muzzle of bear sniffing the bowl for roots and grubs exposed. From the lodgepole forest not thirty yards away came the crack of a branch, and for the longest of moments there was breathless silence, and staring frozen into the dark woods . . .

It has become dreadfully cliché among writers to exclaim how once having met the grizzly on its home turf, one never again looks at the country in the same way. But there are only so many ways to dress up the same naked truth of that statement, and no honest way to omit it, so I'll simply repeat the mantra. Because in truth your eyes will be focused, your ears will be tuned, your nose will be testing the air. Your back will be straighter, your feet will be lighter. One never sleepwalks through grizzlyland, dreaming of other places to be.

Wherever I went that day, exploring miles of what was now grizzly country, all was lit in sharp, crystalline relief. It was an eternal day bathed in the most sparkling sunlight, spent contemplating every track and scat and hank of hair, but also every floral configuration of geranium, valerian, gentian, and lupine. It was a day spent peeking inside the flower head of a dandelion, to find a yellow crab spider gripping an ambushed fly by the head. A gopher tunnel excavated by . . . a bear? Breezes rumbling like gentle thunder in the ears. A tensing whiff of carrion. American kestrels hovering on sharp falcon wings over prairie gardens of scuttling voles and hiding grasshoppers. Here and there were dancing pairs of little black-and-white butterflies spiraling skyward.

Late in the day, I circled back for my last glimpse of grizzly grove and saw above it a golden eagle floating on enormous wings, lazily circling and ascending through a pine green backdrop to the purest blue sky. Nowhere to go, no place to be. White puffclouds building, schooner shadows drifting across the rolling sagebrush sea. "If you can fill the unforgiving minute with sixty seconds' worth of distance run," as Kipling well put it, "Yours is the Earth and everything that's in it." That night at camp, with a hermit thrush singing flute notes over the soothing roar of the river, the longest day of an entire life faded to a close.

I cannot and do not pretend to speak for those few whose lives still carry the daily prospect of disaster at the jaws of professional killers. I cannot speak with the entitlement earned by the Tanzanian farmer who dutifully sleeps among lions, or muster the authority of the Sundarban woodsman who goes to work wearing only faith and a facemask to shield him from tigers. I have only to convey what those of science have found, of the fool's experiment unfolding, and the impending impoverishment of life in the void of great predators. All I can personally but crudely attest is that there is something fundamentally different about a land roamed by big meat-eating beasts, a sense that becomes forcefully apparent in a solitary walk through their realm. And I can only believe, from somewhere deeper than any logic center of the brain, that a life of incomprehensible loneliness awaits a world where the wild things were, but are never to be again.

ACKNOWLEDGMENTS

Many, many thanks to my agent, Russ Galen, who immediately said yes to a wild idea, then worked his magic to find it a warm home. And so also my gratitude to the deciding voices at Bloomsbury USA, who took a chance and let a stranger in the door. And particularly to my editor, Kathy Belden, for her deft and gentle hand.

There are hundreds whose interviews and life histories have given me insights and new appreciations for the wild things, whose proper tributes I can't begin to express in anything short of another book. It has been a privilege and an honor to speak with some of the most brilliant minds and admirable souls in nature.

Among those who took especial pains, Jim Estes graciously sat through many hours of endlessly repeated questions about his life with sea otters. Josh Donlan and Harry Greene, who, like Estes, have suffered no small amount of grief for their ideas, shared a fascinating glimpse into the back rooms and seamy underworld of their profession. Your secrets are safe with me—for this book, anyway.

Josh and Jim, along with Bob Beschta, Donna Hart, Bob Paine, Bill Ripple, Michael Soulé, John Terborgh, and Don Waller each reviewed parts of this manuscript, and each spared me embarrassing gaffes. For those that surely remain, the fault rests here.

A special thanks to Bob Paine, who accompanied me to Holy Ground; to Rodney Bartgis and Melissa Thomas-Van Gundy, who led me to the Hollow;

and to John and Lea Vucetich, who showed me their own Magic Kingdom at Isle Royale.

Thanks also to Kent Redford, for his skeptic's perspective and early warnings of decorator crabs. And to cattleman Jack Turnell, for his stunning candor on wolves and loathing, who with his wife, Lili, graciously invited the enemy to their dinner table.

Martha Rosen, David Steere, and the staff at the Natural History Library in the National Museum of Natural History were forever welcoming and helpful in guiding me through the musty reference stacks and bewildering maze of secret hallways that are a special charm of the Smithsonian.

Scott Weidensaul was throughout a soothing voice of experience, who more than once talked a despondent scribe off the ledge. Trish Tolbert and Will Murray served as mentors and role models, though they probably have no idea. And Pam Davis long ago sacrificed so much to make my leap to the writer's life possible. Thank you.

To the Nature Conservancy and the many admirable colleagues with whom I shared so many life-altering adventures, thanks for the journey. The Conservancy was the mother ship harboring the greatest job in journalism, transporting me to some of the most outrageously wild outposts on the planet. May the good ship keep true to course.

Kathy Kohm at the helm of the magazine *Conservation* (formerly *Conservation in Practice*) provided a welcome forum for exploring certain predator-worthy ideas during the writing of this book. Kathy and her crew also independently came up with the *WTWTW* title before my editors and I did, and generously shared.

From those humbling baptismal days at *Science News*, I owe a long overdue thanks to my editors Laurie Jackson and Pat Young, who caressed and slashed with amazing balance. And to Rich Monastersky, way back "When Life Got Hard," for handing me that predation paper by Vermeij.

To the memory of Laura van Dam, whom I only wish I could thank in person for her utmost kindness and early encouragements.

To Sunshine Annie, DVM, who gave her all to rescue this book from ruin. To the Hairy Legs Ski Team, for that indulgent week of recharge in the

Wasatch powder. And to Jean Rousek, whose glass is not only half full, but somehow forever overflowing—thank you for sharing.

Thanks must go to Meerkat, my faithful desktop editor who supervised many hours of composition, her twitching tail repeatedly clearing the screen of cobwebs.

And to Dee and Kev, Paul and Care, whose selfless heroics on behalf of homeless animals remind me how high my bar has been set.

And most of all to my love Kathy. It is her bottomless compassion for all fellow creatures that gives a wearied chronicler of their destruction his reason for getting up in the morning.

NOTES

Full citations for all references noted herein are found in the bibliography.

Chapter 1: Arms of the Starfish

Pieces of this chapter were published in my 2003 article "Playing God in a Tide Pool." Robert T. Paine's outdoor lab on the rocky coasts of the Pacific Northwest is beautifully portrayed in *The Intertidal Wilderness: A Photographic Journey Through Pacific Coast Tidepools*, by Anne Wertheim Rosenfeld, with contributions from Paine.

The impact of Charles Elton's *Animal Ecology* as a landmark text of ecology (as well as its author as ecological icon) has been nicely interpreted by Mathew A. Leibold and J. Timothy Wootton in their introduction to a 2001 reissue of Elton's 1927 classic by the University of Chicago Press. Elton's role in deciphering population cycles is handily summarized, among other places, in the first chapter of Peter Turchin's book *Complex Population Dynamics* (2003).

Donald Worster's *Nature's Economy* (1994) provides a thorough accounting of the historical context in which Elton's ecology evolved. Several other landmark studies in community ecology and population cycles are to be found in Joseph H. Connell's work on barnacles, Charles J. Krebs and colleagues' long-term studies on the population cycles of lynx and hare, and Raymond L. Lindeman's 1942 paper on the energy flows of ecosystems. For seminal work on predator-prey systems in lakes, see papers by John Langdon Brooks and Stanley I. Dodson, and Stephen B. Carpenter and James F. Kitchell.

Critiques of HSS's green world hypothesis can be found in P. R. Ehrlich and L. C. Birch (1967) and William W. Murdoch (1966), with a more recent overview of the debate by Gary A. Polis (1999).

Timothy Wootton and others among Bob Paine's long and distinguished list of students generously provided insight on the man's character and contributions. The

Importance of Species (2003), edited by Peter Kareiva and Simon A. Levin, serves as both a tribute to Paine and a synthesis on the science of keystone species.

For striking visuals on the intimate act of starfish predation, as well as an extensive overview of the early evolution of life via predation, see the amazing sequences captured by cinematographers in the video documentary *The Shape of Life*, by Ron Bowman and Sea Studios Foundation.

Chapter 2: Planet Predator

Much of the discussion on the genesis of predation and its impact is informed by Stephan Bengston's "Origins and Early Evolution of Predation" (2002). Bits and pieces in the reckoning of the Cambrian explosion were lifted from my 1990 features for *Science News* ("When Life Got Hard") and the *Washington Post* ("Back When Life Got Hard"). Stephen J. Gould's 1989 book, *Wonderful Life*, is a spirited tale of those who've worked to decipher that explosion from the incomparable quarries of the Burgess Shale.

The history of life in North America is admirably synthesized by Tim Flannery in *The Eternal Frontier* (2001). The most recent and ongoing era of its eradication has been well-documented, most gracefully by Peter Matthiessen's *Wildlife in America* (1987). Europe's war with its predators is covered in technical papers by J. C. Reynolds and S. C. Tapper (1996), and Urs Breitenmoser (1998). The plight of America's wolves is particularly well covered by Jon T. Coleman (2004), *Vicious: Wolves and Men in America;* Thomas Dunlap (1988), *Saving America's Wildlife: Ecology and the American Mind, 1850–1990;* and Michael J. Robinson (2005), *Predatory Bureaucracy: The Extermination of Wolves and the Transformation of the West.*

Geerat Vermeij's remarkable life with shells and fascination with predation is engagingly told in his autobiography, *Privileged Hands: A Scientific Life* (1997), as well as in Ron Bowman's PBS documentary *The Shape of Life* (2002).

Chapter 3: Forest of the Sea Otter

Papers by and interviews with James A. Estes inform much of this narrative. Karl W. Kenyon's 1969 monograph on the Aleutian sea otters was, until Estes's body of work, the classic; it is still well worth a read. Richard Ellis provides a more thorough and heartrending account of the Bering Sea slaughters in *The Empty Ocean* (2003).

Chapter 4: The Whale Killer

James Estes and his students are the source for much of this chapter too. Key papers laying out their argument for killer whales as sea otter predators are by Estes et al. (1998), and Terrie M. Williams et al. (2004).

For an astounding accounting of the items known to wind up in the stomachs of killer whales, see Thomas A. Jefferson et al. (1991). For a stunning beachfront seat to killer whales snatching seal pups, see the segment "Coasts," in the incomparable

BBC–Discovery Channel film series, *The Blue Planet* (2001), narrated by David Atten-
borough. In the segment, "Ocean World," the series includes graphic footage of a
pack of killer whales attacking a gray whale mother and calf in Monterey Bay. Viewer
discretion is advised, on both accounts.

The furor over Alan Springer and James Estes's whaling hypothesis (A. M.
Springer et al., 2003) can be appreciated through the raft of critique papers that followed, including those from D. P. DeMaster et al. (2006), S. A. Mizroch and Dale Rice
(2006), and Andrew W. Trites et al. (2007), and climaxing with the 2007 block-
buster by Paul Wade et al. Yet another fascinating take on the mystery, called the scavenging hypothesis, is suggested by Hal Whitehead and Randall Reeves (2005) and
leans favorably toward the Springer and Estes camp. At press time, a reply from
Springer and Estes was under review.

Both sides somehow eventually collaborated for a gigantic and impressive
synthesis on the ecology of whaling in *Whales, Whaling, and Ocean Ecosystems* (James A.
Estes et al., 2006).

Chapter 5: Ecological Meltdown

Much of the bibliographic information on John Terborgh was provided by Terborgh
himself, via interviews and from his writings; in particular, see *Where Have All the Birds
Gone?* (1989), and *Requiem for Nature* (1999). A good deal more was provided by his
sister, Anne Terborgh, and former students and colleagues, most notably Scott
Robinson and David Wilcove.

For a more intimate view of the ecology of Barro Colorado Island—as well as a
differing perspective on the impact of its missing predators—see *A Magic Web*
(2002), by Egbert Giles Leigh Jr., with photographs by Christian Ziegler. For a more
technical counterpoint to Terborgh's interpretation of the predators' role at BCI, see
S. Joseph Wright et al. (1994).

Much of the description of the Guri islands is gleaned and interpreted from interviews with its researchers, John Terborgh, Ken Feeley, and Gabriela Orihuela
chief among them. Visuals and a captivating storyline can be found in the National
Geographic documentary *Strange Days on Planet Earth: Predators* (2005), produced by
Ron Bowman.

Among the reams of publications describing the ecological meltdown of Lago
Guri (see papers by Terborgh and Feeley, among others), Terborgh and colleagues'
2001 *Science* paper most succinctly sums it up, while Jared Diamond's accompanying commentary in the same issue adds color and context.

Chapter 6: Bambi's Revenge

The Science of Overabundance (McShea, Underwood, and Rappole, 1997), though already
slightly outdated, remains a great primer on the history and ecology of too many

deer. Popular accounts include those from Andrew C. Revkin (2002), Erik Ness (2003), and Stephen B. Jones (1993), among many others.

The story of Haida Gwai comes from conversations with Jean-Louis Martin and papers by his crew and colleagues, including those of Sylvain Allombert et al. (2005), Stephen A. Stockton et al. (2005), and Gwenaël Vourc'h et al. (2001–2003). A video documentary handsomely portraying the story is Michel Coqblin and Jean-Louis Martin's *Haïda Gwaii: A Natural Laboratory.*

The deflowering of Cades Cove in the Great Smoky Mountains National Park is documented in papers by Christopher Webster et al. (2005) and Jennifer A. Griggs et al. (2006). Deer problems in the Shenandoahs were informed by conversations with Bill McShea and Chad Stewart at the Smithsonian National Zoological Park's Conservation and Research Center in Front Royal, Virginia, and with Rolf Gubler at Shenandoah National Park. Gary Roisum, manager of Huntley Meadows Park in Alexandria, Virginia, provided a tour through that park's history and ongoing biological invasion and decay.

An impressive body of research on the whitetail's ecological impact can be found at the Web site for Don Waller's lab at the University of Wisconsin-Madison, http://www.botany.wisc.edu/waller/Researchpage.html.

Chapter 7: Little Monsters' Ball

The ecology of mesopredators as presented here was derived in large part through conversations with Stanley D. Gehrt, W. Douglas Robinson, Scott K. Robinson, Kenneth A. Schmidt, Michael E. Soulé, and David S. Wilcove, with accompanying papers from each. Other important authors unmentioned in the text include include Francisco Palomares et al., and Marsha A. Sovada et al. The baboon plague of Africa was informed by Justin Brashares, whose papers were forthcoming, and whose stories as told are more frightening than I have been able to convey.

For cautions on extrapolating the mesopredator release concept too broadly, see Stanley Gehrt and William R. Clark's "Raccoons, Coyotes, and Reflections on the Mesopredator Release Hypothesis" (2003).

The coyote's invasion of eastern and urban America is covered in papers by Matthew E. Gompper (2002) and in a popular account by Mary Battiata (2006), among others.

Extensive summaries on the impact of domestic cats on the world's native wildlife can be found online, in reports by Linda Winter (2006), "Impacts of Feral and Free-ranging Cats on Bird Species of Special Concern," and by John Coleman et al. (1997), "Cats and Wildlife: A Conservation Dilemma." For a sobering look at the cat's ravaging of the world's oceanic island fauna, see Manuel Nogales et al. (2004), "A Review of Feral Cat Eradication on Islands."

Chapter 8: Valley of Fear

William J. Ripple and Robert L. Beschta have contributed the lion's share of published studies in this telling of Yellowstone's ongoing ecological cascade. Their affiliated Web site, www.cof.orst.edu/cascades, is a smorgasbord of their projects and papers, as well as links to some of the endless wave of wolf news emanating from this hotspot of research on terrestrial trophic cascades.

Additional and important contributions on the Yellowstone wolves and elk come from, among others, Eric J. Bergman and colleagues (2006), Hawthorne L. Beyer et al. (2006), Scott Creel et al. (2005), Daniel Fortin et al. (2004, 2005), Jeff P. Hollenbeck and Ripple (2007), Joshua S. Halofsky and Ripple (in press), Matthew J. Kauffman et al. (2007), Julie S. Mao et al. (2005), and John Vucetich et al. (2005).

Yellowstone's history of elk management is harshly, if not savagely, critiqued by Alston Chase (1987) in *Playing God in Yellowstone: The Destruction of America's First National Park*, and by Charles L. Kay (1997) "Viewpoint: Ungulate Herbivory, Willows, and Political Ecology in Yellowstone." The return of Yellowstone's wolves is covered in detail by Hank Fischer (1995) in *Wolf Wars*, and also very stylishly so in Thomas McNamee's (1997) *The Return of the Wolf to Yellowstone*.

The chase scenes were reconstructed from writings and interviews with veteran wolf-watchers Rick McIntyre, Doug Smith, and Daniel MacNulty, and vivid film footage from Bob Landis and Kathryn Pasternak. The elk kill can be seen in Landis and Pasternak's *Wolf Pack*, and the bison battle in their *Thunderbeast*.

Chapter 9: The Lions of Zion

Susan L. Flader's *Thinking like a Mountain: Aldo Leopold and the Evolution of an Ecological Attitude Toward Deer, Wolves and Forests* (1974) informed much of this chapter's perspective of Leopold's relationship with predators, as did William J. Ripple and Robert L. Beschta's "Linking Wolves and Plants: Aldo Leopold on Trophic Cascades" (2005).

An unpublished essay by Berkeley biologist Dale R. McCullough, "Of Paradigms and Philosophies: Aldo Leopold and the Search for a Sustainable Future" provided a warts-and-all examination of the Leopold legend, including an eye-raising hypothesis on his supposed epiphany over the dying wolf that spawned "Thinking like a Mountain."

The life history of the cougar is admirably covered in *Desert Puma*, by Kenneth A. Logan and Linda L. Sweanor.

Chapter 10: Dead Creatures Walking

Parts of this chapter first appeared in the journal *Conservation in Practice* (now named *Conservation*) in my 2006 feature on rewilding, "Where the Wild Things Were," as well as in a 1994 piece in *Nature Conservancy* magazine, "New Views of Ancient Times." Paul Martin's background and early thinking on rewilding is variously documented by himself and others, most recently in his own *Twilight of the Mammoths* (2005), and in Connie Barlow's inspired *Ghosts of Evolution: Nonsensical Fruit, Missing Partners and Other Ecological Anachronisms* (2001).

A sampling of the professional response to C. Josh Donlan et al.'s rewilding paper (2005) includes letters to *Nature* by Martin A. Schlaepfer, Christopher Irwin Smith, Guillaume Chapron, and Eric Dinerstein and W. Robert Irvin. A more thorough critique appears from Dustin R. Rubenstein and colleagues in the journal *Biological Conservation*: "Pleistocene Park: Does Re-wilding North America Represent Sound Conservation for the 21st Century?"

Beyond the accounts in published journals or news reports, the broader public reaction is for bigger books than this. The various samplings of color and profanity included here were provided by Donlan and Greene, from a manuscript in preparation, "NLIMBY: No Lions in My Backyard."

Chapter 11: The Loneliest Predator

George B. Schaller and Gordon R. Lowther's fascinating stint as australopithecine scavengers is covered in Schaller's 1973 book, *Golden Shadows, Flying Hooves*, and in Schaller and Lowther's 1969 technical report "The Relevance of Carnivore Behavior to the Study of Early Hominids."

Discussions on the early evolution of human culture were derived from a variety of sources, most notably from Jared Diamond's *The Third Chimpanzee* (1992) and *Guns, Germs, and Steel* (1997), Edward O. Wilson's *On Human Nature* (1978), Hans Kruuk's *Hunter and Hunted* (2002), Paul Shepard's *Coming Home to the Pleistocene* (1998), and Donna Hart and Robert W. Sussman's *Man the Hunted: Primates, Predators, and Human Evolution* (2005). Certain passages in this chapter were adapted from Stolzenburg (2005), "Where the Wild Things Were."

The portrayal of man-the-runner borrows primarily from David R. Carrier (1984), "The Energetic Paradox of Human Running and Hominid Evolution," and from Dennis M. Bramble and Daniel E. Lieberman (2004), "Endurance Running and the Evolution of Homo."

A physiological explanation for the pronghorn's speed can be found in Stan Lindstedt et al. (1991), "Running Energetics in the Pronghorn Antelope." An evolutionary explanation, involving ghosts, is championed chiefly by John A. Byers in *American Pronghorn: Social Adaptations and the Ghosts of Predators Past* (1997) and more informally in *Built for Speed: A Year in the Life of Pronghorn* (2003). Carol Kaesuk Yoon gives a popular account of Byers's hypothesis in "Pronghorn's Speed May Be Legacy of Past Predators" (1996).

Epilogue: Alone on the Hill

The ongoing onslaught of the sharks, and just a few of the ecological and economic consequences yet discovered, can be found in the 2007 paper by Ransom A. Myers and colleagues, "Cascading Effects of the Loss of Apex Predatory Sharks from a Coastal Ocean." For some recent stats on shark finning, see Shelley Clarke and colleagues (2006), "Global Estimates of Shark Catches Using Trade Records from Commercial Markets."

In addition, a rich vein of new research revealing shark-triggered cascades has lately been coming from the waters of western Australia, described in papers by Michael Heithaus, Aaron Wirsing, Lawrence Dill, and colleagues.

Some of the contradictions of predator control are recently reviewed by me (2006) in "Us or Them," which is pegged to a study by Kim Murray Berger (2006), "Carnivore-Livestock Conflicts."

The coyote's success in the face of an astounding battery of persecution is a remarkable story in itself. For a better sense of the range of technologies employed against coyotes, see Eric M. Gese (2003), "Management of Carnivore Predation as a Means to Reduce Livestock Losses." On a more graphic plane, *Killing Coyote*, by Doug Hawes-Davis, is a fascinating documentary of a venerable Western culture impassioned with seeing this scrappy little carnivore dead. As with the killer whale footage previously noted, some scenes are not for the weak of stomach. And nobody who wants to know something of the character of coyotes and our ambivalence toward them should miss reading J. Frank Dobie's 1949 classic, *The Voice of the Coyote*.

The activities of the U.S. Wildlife Services, which are responsible for managing nuisance wildlife, are published online at: http://www.aphis.usda.gov/wildlife _damage/prog_data_report_FY2006.shtml.

To see the kill statistics, go to http://www.aphis.usda.gov/wildlife_damage/ annual%20tables/Table%20G,%20FY2006.pdf.

A more sensational account of the U.S. Wildlife Services' predatorial practices over the agency's history is provided by Jack Olsen (1974) in *Slaughter the Animals, Poison the Earth*.

Europe's history of persecuting predators is synthesized by Urs Breitenmoser (1991), "Large Predators in the Alps: The Fall and Rise of Man's Competitors," and by J. C. Reynolds and S. C. Tapper (1996): "Control of Mammalian Predators in Game Management and Conservation."

Analyses and elucidations of sprawl and the expanding human footprint, with particular regards to Yellowstone and the Rocky Mountain West, come from papers by Patricia H. Gude, Andrew J. Hansen, and colleagues.

The shifting baselines syndrome was introduced by Daniel Pauly (1995) and elaborated in his 2003 book in collaboration with Jay Maclean: *In a Perfect Ocean*. For further discussions on the syndrome, as well as the rise of slime, read Jeremy Jackson's 2004 speech "Habitat Destruction and Ecological Extinction of Marine Invertebrates." And for a deflating look at the state of the ocean's fishes, see Jackson and colleagues' "Historical Overfishing and the Recent Collapse of Coastal Ecosystems" (2001), as well as Ransom Myers and Boris Worm's (2003) report in *Nature*, "Rapid Worldwide Depletion of Predatory Fish Communities."

A masterful exploration of man's mental affair with several of his most dangerous predators is David Quammen's *Monster of God* (2003). A more tightly focused story of one community's dilemma dealing with mountain lions in their midst is well told by David Baron in *The Beast in the Garden* (2003). Our age-old fascination with wolves is elegantly portrayed by Barry Lopez in *Of Wolves and Men* (1978). And the grizzly bear's allure can be better appreciated by reading *The Great Bear* (1992), a collection of essays by a range of accomplished writers, edited by John A. Murray. Or better yet, as I've mentioned a time or two, by meeting the great bear on its home turf.

BIBLIOGRAPHY

Abrams, Peter A. 1994. The fallacies of "ratio-dependent" predation. *Ecology* 75(6):1842–50.

———. 2000. The evolution of predator-prey interactions: Theory and evidence. *Annual Review of Ecology and Systematics* 31:79–105.

Abrams, Peter A., and Lev R. Ginzburg. 2000. The nature of predation: Prey dependent, ratio dependent or neither? *Trends in Ecology and Evolution* 15(8):337–41.

Adams, Charles C. 1925. The conservation of predatory mammals. *Journal of Mammalogy* 6(2):83–96.

Adams, D. B. 1979. The cheetah: Native American. *Science* 205:1155–58.

Alderman, Jesse Harlan. 2007. Idaho governor calls for gray wolf kill. *Idaho Statesman*, January 11.

Allen, Durward L. 1974. *Our wildlife legacy*. New York: Funk and Wagnalls.

Allombert, Sylvain, Anthony J. Gaston, and Jean-Louis Martin. 2005. A natural experiment on the impact of overabundant deer on songbird populations. *Biological Conservation* 126:1–13.

Allombert, Sylvain, Steve Stockton, and Jean-Louis Martin. 2005. A natural experiment on the impact of overabundant deer on forest invertebrates. *Conservation Biology* 19(6):1917–29.

Alroy, John. 2001. A multispecies overkill simulation of the end-Pleistocene megafaunal mass extinction. *Science* 292:1893–96.

Alverson, William S., Walter Kuhlmann, and Donald M. Waller. 1994. *Wild forests: Conservation biology and public policy*. Washington, D.C.: Island Press.

Alverson, William S., and Donald M. Waller. 1997. Deer populations and the widespread failure of hemlock regeneration in northern forests. In *The science of overabundance: Deer ecology and population management*, ed. William J. McShea, H. Brian Underwood, and John H. Rappole, 280–97.

Alverson, William S., Donald M. Waller, and Stephen L. Solheim. 1988. Forests too deer: Edge effects in Northern Wisconsin. *Conservation Biology* 2(4):348–58.

Andelman, Sandy J., and William F. Fagan. 2000. Umbrellas and flagships: Efficient conservation surrogates or expensive mistakes? Proceedings of the National Academy of Sciences 97(11):5954–59.

Ardrey, Robert. 1968. African genesis. New York: Dell Publishing.

Arthur, L. M. 1981. Coyote control: The public response. Journal of Range Management 34:14–15.

Asquith, Nigel M., S. Joseph Wright, and Maria J. Clauss. 1997. Does mammal community composition control recruitment in neotropical forests? Evidence from Panama. Ecology 78(3):941–46.

Associated Press. 2004. Study: Bears kill more elk than wolves. New York Times, November 2.

———. 2005. Lions and elephants on the Great Plains? Scientists suggest relocating African species to North America. http://www.cnn.worldnews.com (accessed August 23, 2005).

———. 2007. Wyo lawmakers express wolf plan doubts. CasperStarTribune.net.

Avril, Tom. 2005. Scientists say deer hunting must increase. Philadelphia Inquirer, January 12.

Bailey, Vernon, Joseph Dixon, E. A. Goldman, Edmund Heller, and Charles C. Adams. 1928. Report of the committee on wildlife sanctuaries, including provision for predatory mammals. Journal of Mammalogy 9(4):354–58.

Baker, Bruce W., Heather C. Ducharme, David C. S. Mitchell, Thomas R. Stanley, and H. Paul Peinetti. 2005. Interaction of beaver and elk herbivory reduces standing crop of willow. Ecological Applications 15(1):110–18.

Baker, Philip J., Amy J. Bentley, Rachel J. Ansell, and Stephen Harris. 2005. Impact of predation by domestic cats Felis catus in an urban area. Mammal Review 35:302–12.

Balcomb, Ken. 1985. Cooperative killers. In All the world's animals: Sea mammals, ed. David Macdonald, 38–39.

Ballance, Lisa T., Robert L. Pitman, Roger P. Hewitt, Donald B. Siniff, Wayne Z. Trivelpiece, Phillip J. Clapham, and Robert L. Brownell. 2006. The removal of large whales from the Southern Ocean. In Whales, whaling, and ocean ecosystems, ed. James A. Estes, Douglas P. Demaster, Daniel F. Doak, Terrie M. Williams, and Robert L. Brownell Jr., 215–30.

Balukjian, Brad. 2002. Islands on the brink. Duke Magazine, January–February.

Bangs, Ed, Joe Fontaine, Mike Jimenez, Tom Meier, Carter Niemeyer, Doug Smith, Kerry Murphy, Deb Guernsey, Larry Handegard, Mark Collinge, Rod Krischke, John Shivik, Curt Mack, Issac Babcock, Val Asher, and Dominic Domenici. 2001. Gray wolf restoration in the Northwestern United States. Endangered Species Update 18(4):147–52.

Banta, Joshua A., Alejandro A. Royo, Chad Kirschbaum, and Walter P. Carson. 2005. Plant communities growing on boulders in the Allegheny National Forest: Evidence for boulders as refugia from deer and as a bioassay of overbrowsing. Natural Areas Journal 25(1):10–18.

Barker, Rocky. 2007. Wolf rally draws a crowd. Idaho Statesman, January 12.

Barlow, Connie. 1999. Rewilding for evolution. Wild Earth, Spring: 53–56.

———. 2001. Ghosts of evolution: Nonsensical fruit, missing partners and other ecological anachronisms. New York: Basic Books.

Barnett, Ross, Nobuyuki Yamaguchi, Ian Barnes, and Alan Cooper. 2006. Lost populations and preserving genetic diversity in the lion *Panthera leo*: Implications for its *ex situ* conservation. *Conservation Genetics*, doi: 10.1007/s10592-005-9062-0.

Barnosky, Anthony D., Paul L. Koch, Robert S. Feranec, Scott L. Wing, and Alan B. Shabel. 2004. Assessing the causes of late Pleistocene extinctions on the continents. *Science* 306:70–75.

Baron, David. 2003. *The beast in the garden: A modern parable of man and nature.* New York: W. W. Norton and Company.

Barrett-Lennard, Lance G., Kathy Heise, Eva Saulitis, Graeme Ellis, and Craig Matkin. 1995. *The impact of killer whale predation on Steller sea lion populations in British Columbia and Alaska.* Report for the North Pacific Universities Marine Mammal Research Consortium Fisheries Centre, University of British Columbia.

Barryman, Alan A. 1992. The origins and evolution of predator-prey theory. *Ecology* 73:1529–39.

Bascompte, Jordi, Carlos J. Melián, and Enric Sala. 2005. Interaction strength combinations and the overfishing of a marine food web. *Proceedings of the National Academy of Sciences* 102(15):5443–47.

Basgall, Monte. 2002. Identifying the forests' prime evil. *Duke Magazine*, January–February.

Battiata, Mary. 2006. Among us. *Washington Post Magazine*, April 16: 6–21.

Baum, Julia K., and Ransom A. Myers. 2004. Shifting baselines and the decline of pelagic sharks in the Gulf of Mexico. *Ecology Letters* 7:135–45.

Baum, Julia K., Ransom A. Myers, Daniel G. Kehler, Boris Worm, Shelton J. Harley, and Penny A. Doherty. 2003. Collapse and conservation of shark populations in the Northwest Atlantic. *Science* 299:389–92.

Beier, Paul. 1991. Cougar attacks on human in the United States and Canada. *Wildlife Society Bulletin* 19:403–412.

Bengeyfield, Pete. 2003. *Incredible vision: The wildlands of Greater Yellowstone.* Helena, MT: Riverbend Publishing.

Bengston, Stephan. 2002. Origins and early evolution of predation. In *Paleontological Society Special Papers* 8, ed. M. Kowalewski and P. H. Kelley, 289–318.

Berger, Joel. 1979. "Predator harassment" as a defensive strategy in ungulates. *American Midland Naturalist* 102(1):197–99.

———. 1998. Future prey: Some consequences of losing and restoring large carnivores. In *Behavioral ecology and conservation biology*, ed. T. Caro, 80–100. Oxford: Oxford University Press.

———. 1999. Anthropogenic extinction of top carnivores and interspecific animal behaviour: Implications of the rapid decoupling of a web involving wolves, bears, moose and ravens. *Proceedings of the Royal Society* 266:2261–67.

———. 2002. Wolves, landscapes, and the ecological recovery of Yellowstone. *Wild Earth*, Spring: 32–37.

————. 2005. Hunting by carnivores and by humans: Is functional redundancy possible and who really cares? In *Large carnivores and the conservation of biodiversity*, ed. Justina Ray, Kent H. Redford, Robert S. Steneck, and Joel Berger, 316–41.

————. 2007. Carnivore repatriation and Holarctic prey: Narrowing the deficit in ecological effectiveness. *Conservation Biology*. 21(4):1105–16.

Berger, Joel, and Douglas Smith. 2005. Restoring functionality in Yellowstone with recovering carnivores: Gains and uncertainties. In *Large carnivores and conservation of biodiversity*, ed. Justina Ray, Kent H. Redford, Robert S. Steneck, and Joel Berger, 100–9.

Berger, Joel, Peter B. Stacey, Lori Bellis, and Matthew P. Johnson. 2001. A mammalian predator-prey imbalance: Grizzly bear and wolf extinction affect avian neotropical migrants. *Ecological Applications* 11:947–60.

Berger, Joel, Jon E. Swenson, and Inga-Lill Persson. 2001. Recolonizing carnivores and naïve prey: Conservation lessons from Pleistocene extinctions. *Science* 291:1036–39.

Berger, Joel, and John D. Wehausen. 1991. Consequences of a mammalian predator-prey disequilibrium in the Great Basin Desert. *Conservation Biology* 5(2):244–48.

Berger, Kim Murray. 2006. Carnivore-livestock conflicts: Effects of subsidized predator control and economic correlates on the sheep industry. *Conservation Biology* 20(3):751–61.

Berger, Lee R. 2006. Brief communication: Predatory bird damage to the Taung type-skull of *Australopithecus africanus* Dart. 1925. *American Journal of Physical Anthropology* 131:166–68.

Berger, L. R., and R. J. Clarke. 1995. Eagle involvement in accumulation of the Taung child fauna. *Journal of Human Evolution* 29:275–99.

Bergman, Eric J., Robert A. Garrott, Scott Creel, John J. Borkowski, Rosemary Jaffe, and F. G. R. Watson. 2006. Assessment of prey vulnerability through analysis of wolf movements and kill sites. *Ecological Applications* 16:273–84.

Beschta, Robert L. 2003. Cottonwoods, elk, and wolves in the Lamar Valley of Yellowstone National Park. *Ecological Applications* 13:1295–1309.

————. 2005. Reduced cottonwood recruitment following extirpation of wolves in Yellowstone's northern range. *Ecology* 86:391–403.

Beschta, Robert L., and William J. Ripple. 2006. River channel dynamics following extirpation of wolves in northwestern Yellowstone National Park. *Earth Surface Processes and Landforms* 31:1525–39.

————. 2007. Wolves, elk, and aspen in the winter range of Jasper National Park, Canada. *Canadian Journal of Forest Resources* 37:1–13.

————. In press. Increased willow heights along northern Yellowstone's Blacktail Deer Creek following wolf reintroduction. *Western North American Naturalist*.

Beyer, Hawthorne L. 2006. Wolves, elk and willow on Yellowstone National Park's northern range. M.S. thesis, University of Alberta, Edmonton, Alberta.

Beyer, Hawthorne L., Evelyn H. Merrill, Nathan Varley, and Mark S. Boyce. 2007. Willow on Yellowstone's northern range: Evidence for a trophic cascade in a large mammalian predator-prey system? *Ecological Applications* 17(6):1563–71.

Bildstein, Keith L. 2001. Raptors as vermin: A history of human attitudes towards Pennsylvania's birds of prey. *Endangered Species Update* 18(4):124–28.

Binder, W. J., and B. Van Valkenburgh. 2000. Development of bite strength and feeding behaviour in spotted hyaenas (*Crocuta crocuta*). *Journal of Zoology* 252:273–83.

Binkley, Dan, Margaret M. Moore, William H. Romme, and Peter M. Brown. 2006. Was Aldo Leopold right about the Kaibab deer herd? *Ecosystems* 9:227–41.

Birkeland, Charles, and Paul K. Dayton. 2005. The importance in fishery management of leaving the big ones. *Trends in Ecology and Evolution* 20(7):356–58.

Bjorndal, Karen A., and Alan B. Bolten. 2003. From ghosts to key species: Restoring sea turtle populations to fulfill their ecological roles. *Marine Turtle Newsletter* 100:16–21.

Blejwas, Karen M., Benjamin N. Sacks, Michael M. Jaeger, and Dale R. McCullough. 2002. The effectiveness of selective removal of breeding coyotes in reducing sheep predation. *Journal of Wildlife Management* 66(2):451–62.

Block, Barbara A., Steven L. H. Teo, Andreas Walli, Andre Boustany, Michael J. W. Stokesbury, Charles J. Farwell, Kevin C. Weng, Heidi Dewar, and Thomas D. Williams. 2005. Electronic tagging and population structure of Atlantic bluefin tuna. *Nature* 434:1121–26.

Blumenschine, Robert J. 1991. Hominid carnivory and foraging strategies, and the socioeconomic function of early archaeological sites. *Philosophical Transactions of the Royal Society of London* 334:211–21.

Blumenschine, Robert J., and John A. Cavallo. 1992. Scavenging and human evolution. *Scientific American*, October: 90–96.

Boitani, Luigi. 2003. Wolf conservation and recovery. In *Wolves: Behavior, ecology, and conservation*, ed. L. David Mech and Luigi Boitani, 317–40.

Bolger, Douglas T., Allison C. Alberts, and Michael E. Soulé. 1991. Occurrence of bird species in habitat fragments: Sampling, extinction, and nested species subsets. *American Naturalist* 137(2):155–66.

Bosire, Bogonko. 2005. African conservationists denounce proposal for giant US wildlife park. *Terra Daily*, August 18. http://www.terradaily.com/news/life-05zzzx.html (accessed August 18, 2005).

Bowman, Ron, producer. 2002. *The shape of life: The survival game*, episode 6 of an eight-part video series by Sea Studios Foundation.

———. 2005. *Strange days on planet Earth: Predators*, episode 3 of a four-part video series by National Geographic.

Bramble, Dennis M., and Daniel E. Lieberman. 2004. Endurance running and the evolution of Homo. *Nature* 432:345–52.

Branch, Trevor A., and Terrie M. Williams. 2006. Legacy of industrial whaling: Could killer whales be responsible for declines of sea lions, elephant seals, and minke whales in the southern hemisphere? In *Whales, whaling, and ocean ecosystems*, ed. James A. Estes, Douglas P. Demaster, Daniel F. Doak, Terrie M. Williams, and Robert L. Brownell Jr., 262–78.

Breitenmoser, Urs. 1991. The reintroduction of the lynx in Switzerland. In *Great cats: Majestic creatures of the wild*, ed. John A. Seidensticker and Susan Lumpkin, 230.

———. 1998. Large predators in the Alps: The fall and rise of man's competitors. *Biological Conservation* 83:279–89.

Briggs, Derek E. G., and Harry B. Whittington. 1985. Terror of the trilobites. *Natural History* 12:34–39.

Brooks, John Langdon, and Stanley I. Dodson. 1965. Predation, body size, and the composition of plankton. *Science* 150:28–35.

Brown, David E., ed. 1983. *The wolf in the Southwest: The making of an endangered species*. Tucson: University of Arizona Press.

Brown, James H., and Edward J. Heske. 1990. Control of a desert-grassland transition by a keystone rodent guild. *Science* 250:1705–7.

Brown, Janet, and Michael Neve, eds. 1989. *Charles Darwin: Voyage of the Beagle. Charles Darwin's Journal of Researches*. London: Penguin Books.

Brown, Joel S., John W. Laundré, and Mahesh Gurung. 1999. The ecology of fear: Optimal foraging, game theory, and trophic interactions. *Journal of Mammalogy* 80:385–99.

Buchmann, Stephen L., and Gary Paul Nabhan. 1996. *The forgotten pollinators*. Washington, D.C.: Island Press.

Budiansky, Stephen. 1994. Deer, deer everywhere. *U.S. News and World Report*, November 21: 85–86.

———. 1996. Yellowstone's unraveling. *U.S. News and World Report*, September 16: 80–83.

Burk, J. C. 1973. The Kaibab deer incident: A long-persisting myth. *BioScience* 9:43–47.

Burney, David A., and Timothy F. Flannery. 2006. Fifty millennia of catastrophic extinctions after human contact. *Trends in Ecology and Evolution* 20(7):395–401.

Butchko, Peter H. 1990. Predator control for the protection of endangered species in California. In *Proceedings of the Fourteenth Vertebrate Pest Conference, Davis, California*, ed. J. E. Borrecco and R. E. Marsh, 237–40. Davis: University of California.

Byatt, Andy, producer. 2001. *The blue planet: Open ocean*, episode 3 of an eight-part film series by BBC/Discovery Channel.

Byers, John A. 1997. *American pronghorn: Social adaptations and the ghosts of predators past*. Chicago: University of Chicago Press.

———. 2003. *Built for speed: A year in the life of pronghorn*. Cambridge, MA: Harvard University Press.

Cabeza, Mar, Anni Arponen, and Astrid van Teeffelen. 2007. Top predators: Hot or not? A call for systematic assessment of biodiversity surrogates. *Journal of Applied Ecology*, doi: 10.1111/j.1365-2664.2007.01364.x.

Cahalane, Victor H. 1961. *Mammals of North America*. New York: Macmillan.

Callicott, J. Baird. 2002. Choosing appropriate temporal and spatial scales for ecological restoration. *Journal of Bioscience* 27(4):409–20.

Canisius College. 2006. Crafty killer whales demonstrate "cultural learning." Press release, January 25.

Caras, Roger. 1996. *A perfect harmony: The intertwining lives of animals and humans throughout history.* New York: Simon & Schuster.

Cardillo, Marcel, Georgina M. Mace, John L. Gittleman, and Andy Purvis. 2006. Latent extinction risk and the future battlegrounds of mammal conservation. *Proceedings of the National Academy of Sciences* 103(11):4157–61.

Cardillo, Marcel, Andy Purvis, Wes Sechrest, John L. Gittleman, Jon Bielby, and Georgina M. Mace. 2004. Human population density and extinction risk in the world's carnivores. *PLoS Biology* 2(7):0909–14.

Carey, John. 1987. Who's afraid of the big bad wolf? *National Wildlife*, August–September: 4–10.

Caro, Tim. 2000. Focal species. *Conservation Biology* 14(6):1569–70.

Carpenter, Stephen R., and James F. Kitchell. 1988. Consumer control in lake productivity. *BioScience* 38:764–69.

Carpenter, Stephen R., James F. Kitchell, and James R. Hodgson. 1985. Cascading trophic interactions and lake productivity. *BioScience* 35:634–39.

Carrier, David R. 1984. The energetic paradox of human running and hominid evolution. *Current Anthropology* 25(4):483–95.

Carson, Rachel. 1962. *Silent Spring.* New York: Fawcett World Library.

Cartmill, Matt. 1993. The Bambi syndrome. *Natural History*, June: 6–12.

Caughley, G. 1970. Eruption of ungulate populations, with emphasis on Himalayan thar in New Zealand. *Ecology* 51:53–71.

Chadwick, Douglas H. 1998. Return of the gray wolf. *National Geographic*, May: 72–99.

Chapron, Guillaume. 2005. Re-wilding: Other projects help carnivores stay wild. *Nature* 437:318.

Chase, Alston. 1987. *Playing God in Yellowstone: The destruction of America's first National Park.* San Diego: Harcourt Brace and Company.

Chase, Jonathan M. 2000. Are there real differences among aquatic and terrestrial food webs? *Trends in Ecology and Evolution* 15(10):408–12.

Chase, Jonathan M., Peter A. Abrams, James P. Grover, Sebastian Diehl, Peter Chesson, Robert D. Holt, Shane A. Richards, Robert M. Nisbet, and Ted J. Case. 2002. The interaction between predation and competition: A review and synthesis. *Ecology Letters* 5:302–15.

Chavez, Andreas S., Eric M. Gese, and Richard S. Krannich. 2005. Attitudes of rural landowners toward wolves in northwestern Minnesota. *Wildlife Society Bulletin* 33(2):517–27.

Chew, Ryan. 2005. Stakeout: A tale of two species. *Chicago Wilderness*, Fall. http://chicagowildernessmag.org/issues/fall2005/stakeout.html (accessed December 21, 2007).

Chinery, Michael. 1990. *Predators: Killers of the wild.* London: Bedford Editions.

Christian Science Monitor. 2005. Rewilding America, Pleistocene style. August 30.

Christiansen, Per. 2002. Locomotion in terrestrial mammals: The influence of body mass, limb length and bone proportions on speed. *Zoological Journal of the Linnean Society* 136(4):685–714.

Clapham, Phillip J., and Jason S. Link. 2006. Whales, whaling, and ecosystems in the North Atlantic Ocean. In *Whales, whaling, and ocean ecosystems*, ed. James A. Estes, Douglas P. Demaster, Daniel F. Doak, Terrie M. Williams, and Robert L. Brownell Jr., 314–23.

Clark, Tim W., A. Peyton Curlee, Steven C. Minta, and Peter M. Kareiva. 2000. *Carnivores in ecosystems: The Yellowstone experience.* New Haven, CT: Yale University Press.

Clarke, Shelley, Murdoch K. McAllister, E. J. Milner-Gulland, G. P. Kirkwood, Catherine G. J. Michielsen, David J. Agnew, Ellen K. Pikitch, Hidecki Nakano, Mahmood S. Shivji. 2006. Global estimates of shark catches using trade records from commercial markets. *Ecology Letters* 9:1115–26.

Colbert, E. H. 1980. *Evolution of the vertebrates.* 3rd ed. New York: John Wiley.

Coleman, Jon T. 2004. *Vicious: Wolves and men in America.* New Haven, CT: Yale University Press.

Coleman, John, Stanley Temple, and Scott Craven. 1997. Cats and wildlife: A conservation dilemma. http://www.cnr.vt.edu/extension/fiw/wildlife/damage/Cats.pdf (accessed December 21, 2007).

Colinvaux, Paul. 1978. *Why big fierce animals are rare: An ecologist's perspective.* Princeton, NJ: Princeton University Press.

Comisky, Lauren, Alejandro A. Royo, and Walter P. Carson. 2005. Deer browsing creates rock refugia gardens on large boulders in the Allegheny National Forest, Pennsylvania. *American Midland Naturalist* 154:201–6.

Connell, Joseph H. 1961a. Effect of competition, predation by *Thais lapillus*, and other factors on natural populations of the barnacle *Balanus balanoides*. *Ecological Monographs* 31(1):61–104.

———. 1961b. The influence of interspecific competition and other factors on the distribution of the barnacle *Chthamalus stellatus*. *Ecology* 42(4):710–23.

Coope, G. R. 1995. Insect faunas in ice age environments: Why so little extinction? In *Extinction rates*, ed. J. H. Lawton and R. M. May 55–74. Oxford: Oxford University Press.

Coqblin, Michel, and Jean-Louis Martin. 2003. *Haïda Gwaii: A natural laboratory.* Mille et Une Productions / CNRS Images.

Côté, Steeve D. 2005. Extirpation of a large black bear population by introduced white-tailed deer. *Conservation Biology* 19(5):1668–71.

Côté, Steeve D., Thomas P. Rooney, Jean-Pierre Tremblay, Christian Dussalt, and Donald M. Waller. 2004. Ecological impacts of deer overabundance. *Annual Review of Ecology, Evolution, and Systematics* 35:113–47.

Courchamp, Franck, Michel Langlais, and George Sugihara. 1999. Cats protecting birds: Modelling the mesopredator release effect. *Journal of Animal Ecology* 68:282–92.

Creel, Scott, and John A. Winnie Jr. 2005. Responses of elk herd size to fine-scale spatial and temporal variation in the risk of predation by wolves. *Animal Behaviour* 69:1181–89.

Creel, Scott, John Winnie Jr., Bruce Maxwell, Ken Hamlin, and Michael Creel. 2005. Elk alter habitat selection as an antipredator response to wolves. *Ecology* 86(12):3387–97.

Crête, Michel. 1999. The distribution of deer biomass in North America supports the hypothesis of exploitation ecosystems. *Ecology Letters* 2:223–27.

Crête, Michel, and Micheline Manseau. 1996. Natural regulation of cervidae along a 1000 km latitudinal gradient: Change in trophic dominance. *Evolutionary Ecology* 10:51–62.

Croker, Robert A. 1991. *Pioneer ecologist: The life and work of Victor Ernest Shelford 1877–1968.* Washington, D.C.: Smithsonian Institution Press.

Croll, Donald A., Raphael Kudela, and Bernie R. Tershy. 2006. Ecosystem impact of the decline of large whales in the North Pacific. In *Whales, whaling, and ocean ecosystems*, ed. James A. Estes, Douglas P. Demaster, Daniel F. Doak, Terrie M. Williams, and Robert L. Brownell Jr., 202–14.

Crooks, Kevin R., and Michael E. Soulé. 1999. Mesopredator release and avifaunal extinctions in a fragmented system. *Nature* 400:563–66.

Crowcroft, Peter. 1991. *Elton's ecologists: A history of the Bureau of Animal Population.* Chicago: University of Chicago Press.

Curry, Jessica. 2005. The evolution of the urban coyote: The daily routine of a predator-hunter revolves around odor-free existence. *Chicago Life*, November.

Curtis, John T. 1959. *The vegetation of Wisconsin.* Madison: University of Wisconsin Press.

Dahlheim, Marilyn E., and John E. Heyning. 1999. Killer whale *Orcinus orca* (Linnaeus, 1758). In *Handbook of Marine Mammals* 6, ed. Sam H. Ridgway and Sir Richard Harrison, 281–321.

Dalton, Rex. 2005. Is this any way to save a species? *Nature* 436:14–16.

Danner, Eric M., Matthew J. Kauffman, and Robert L. Brownell. 2006. Industrial whaling in the North Pacific Ocean 1952–1978: Spatial patterns of harvest and decline. In *Whales, whaling, and ocean ecosystems*, ed. James A. Estes, Douglas P. DeMaster, Daniel F. Doak, Terrie M. Williams, and Robert L. Brownell Jr., 134–44.

Darwin, Charles. 1939. *Journal of researches into the geology and natural history of the various countries visited by H.M.S. Beagle.* London: Henry Colburn.

———. 1958. *The origin of species by means of natural selection or the preservation of favoured races in the struggle for life.* Mentor ed. New York: New American Library.

Dawkins, R., and J. R. Krebs. 1979. Arms races between and within species. *Proceedings of the Royal Society of London B* 205:489–511.

Dawson, Pat. 1992. Wolf hearing marked by howls and snarls. *High Country News*, September 7: 3.

Dayton, Leigh. 2001. Mass extinctions pinned on ice age hunters. *Science* 292:1819.

DeByle, Norbert V., and Robert P. Winokur, eds. 1985. *Aspen: Ecology and management in the western United States.* USDA Forest Service General Technical Report RM-119.

deCalesta, David. 1997. Deer and ecosystem management. In *The science of overabundance: Deer ecology and population management*, ed. William J. McShea, H. Brian Underwood, and John H. Rappole, 267–79.

Deeke, Volker B., John K. B. Ford, and Peter J. B. Slater. 2005. The vocal behaviour of mammal-eating killer whales: Communicating with costly calls. *Animal Behaviour* 69:395–405.

Defenders of Wildlife. 2006. New successes in nonlethal wolf control lead to zero wolf-related livestock losses for local ranchers. http://www.defenders.org/newsroom/

press_releases_folder/2006/10_25_2006_successes_in_nonlethal_wolf_control.php ?ht=range%2oriders%2orange%2oriders (accessed September 14, 2007).

DeMaster, D. P., A. W. Trites, P. Clapham, S. Mizroch, P. Wade, R. J. Small, and J. V. Hoef. 2006. The sequential megafaunal collapse hypothesis: Testing with existing data. *Progress in Oceanography* 68:329–42.

Denevan, William M. 1992. The pristine myth: The landscape of the Americas in 1492. *Annals of the Association of American Geographers* 82(3):369–85.

Derocher, Andrew E., Nicholas J. Lunn, and Ian Stirling. 2004. Polar bears in a warming climate. *Integrative and Comparative Biology* 44(2):163–76.

Diamond, Jared. 1987. The worst mistake in the history of the human race. *Discover*, May: 64–66.

———. 1989. Quaternary megafaunal extinctions: Variations on a theme by Paganini. *Journal of Archaeological Science* 16:167–75.

———. 1992a. Must we shoot deer to save nature? *Natural History* 8:2–8.

———. 1992b. *The third chimpanzee: The evolution and future of the human animal.* New York: Harper-Collins Publishers.

———. 1997. *Guns, germs, and steel: The fates of human societies.* New York: W. W. Norton and Company.

———. 2001. Dammed experiments! *Science* 294:1847–48.

———. 2002. Evolution, consequences and future of plant and animal domestication. *Nature* 418:700–7.

Dice, Lee R. 1925. Scientific value of predatory mammals. *Journal of Mammalogy* 6(1):25–27.

Dietl, Gregory P. 2003. The escalation hypothesis: One long argument. *Palois* 18:83–86.

Dietl, Gregory P., Gregory S. Herbert, and Geerat J. Vermeij. 2004. Reduced competition and altered feeding behavior among marine snails after a mass extinction. *Science* 306:2229–31.

Dill, Lawrence M., Michael R. Heithaus, and Carl J. Walters. 2003. Behaviorally mediated indirect interactions in marine communities and their conservation and implications. *Ecology* 84(5):1151–57.

Dinerstein, Eric, and W. Robert Irvin. 2005. Re-wilding: No need for exotics as natives return. *Nature* 437:476.

Di Silvestro, Roger. 2007. Fair funding for endangered species. *National Wildlife.* http://www.nwf.org/nationalwildlife/article.cfm?issueID=116&articleID=1488 (accessed December 21, 2007).

———. 2007. Fair funding for wildlife: Investing in our commitment to save America's endangered wildlife. A report by the National Wildlife Federation.

Doak, Daniel, Terrie M. Williams, and James A. Estes. 2006. In *Whales, whaling, and ocean ecosystems.* ed. James A. Estes, Douglas P. DeMaster, Daniel F. Doak, Terrie M. Williams, and Robert L. Brownell Jr., 231–44.

Dobie, J. Frank. 1949. *The voice of the coyote.* Lincoln: University of Nebraska Press.

Dobson, Andy, Isabella Cattadori, Robert D. Holt, Richard S. Ostfeld, Felicia Keesing, Kristle

Krichbaum, Jason R. Rohr, Sarah E. Perkins, Peter J. Hudson. 2006. Sacred cows and sympathetic squirrels: The importance of biological diversity to human health. PLoS Medicine 3(6):e231, doi: 10.1371/journal.pmed.0030231.

Dobson, M. 1994. Patterns of distribution in Japanese land mammals. Mammal Review 24:91–111.

Domico, Terry. 1988. Bears of the world. New York: Facts on File.

Donlan, C. Josh. 2005a. Claws and effects. How a plan to return big beasts to North America raised hackles and hopes. Grist, November 8.

———. 2005b. Lions and cheetahs and elephants, oh my! Slate, August 18.

Donlan, C. Josh, and Tosha Comendant. 2003. Getting rid of rats. E/The Environmental Magazine, September–October: 10.

Donlan, Josh, Harry W. Greene, Joel Berger, Carl E. Bock, Jane H. Bock, David A. Burney, James A. Estes, Dave Forman, Paul S. Martin, Gary W. Roemer, Felisa A. Smith, and Michael E. Soulé. 2005. Re-wilding North America. Nature 436:913–14.

Donlan, C. Josh., and Paul S. Martin. 2003. Role of ecological history in invasive species management and conservation. Conservation Biology 18(1):267–69.

Doroff, A. M., J. A. Estes, M. T. Tinker, D. M. Burn, and T. J. Evans. 2003. Sea otter population declines in the Aleutian archipelago. Journal of Mammalogy 84:55–64.

Downes, Lawrence. 2005. The shy, egg-stealing neighbor you didn't know you had. New York Times, November 6.

Drayton, Brian, and Richard B. Primack. 1996. Plant species lost in an isolated conservation area in metropolitan Boston from 1894 to 1993. Conservation Biology 10(1):30–39.

Duffy, J. Emmett. 2002. Biodiversity and ecosystem function: The consumer connection. Oikos 99:201–219.

———. 2003. Biodiversity loss, trophic skew and ecosystem functioning. Ecology Letters 6:680–87.

Duggins, David O. 1980. Kelp beds and sea otters: An experimental approach. Ecology 61:447–53.

Dunlap, Thomas. 1988. Saving America's wildlife: Ecology and the American mind, 1850–1990. Princeton, NJ: Princeton University Press.

Dutton, C. E. 1882. The physical geology of the Grand Canyon district. U.S. Geological Survey, 2nd Annual Report, 49–166.

Eason, P. 1989. Harpy eagle attempts predation on adult howler monkey. Condor 91(2):469–70.

Ehrlich, P. R., and L. C. Birch. 1967. The "balance of nature" and "population control." American Naturalist 101:97–107.

Ellis, Richard. 1975. The book of sharks. New York: Grosset and Dunlap.

———. 2003. The empty ocean: Plundering the world's marine life. Washington, D.C.: Island Press.

Ellison, Aaron M. 2006. What makes an ecological icon? Bulletin of the Ecological Society of America, October: 380–86.

Elton, Charles S. 1942. Voles, mice and lemmings: Problems in population dynamics. Oxford: Clarendon Press.

————. 1966. *The pattern of animal communities*. London: Methuen and Co.

————. 1968. *The ecology of animals*. London: Methuen and Co.

————. 2000. *The ecology of invasions by animals and plants*. Chicago: University of Chicago Press.

————. 2001. *Animal ecology*. Chicago: University of Chicago Press.

Elton, Charles, and Mary Nicholson. 1942. The ten-year cycle of the lynx in Canada. *Journal of Animal Ecology* 11(2):215–44.

Emmons, Louise H. 1987. Comparative feeding ecology of felids in a neotropical rainforest. *Behavioral Ecology and Sociobiology* 20:271–83.

Environmental News Network. 1998. Killer whales put Alaska sea otters at risk. October 16.

Erickson, Gregory M., Samuel D. Van Kirk, Jinntung Su, Marc E. Levenston, William E. Caler, and Dennis R. Carter. 1996. Bite-force estimation for *Tyrannosaurus rex* from tooth-marked bones. *Nature* 382: 706–8.

Erlinge, Sam, Görgen Göranson, Göran Högstedt, Göran Jansson, Olof Liberg, Jon Loman, Ingvar N. Nilsson, Torbjörn von Schantz, and Magnus Sylvén. 1984. Can vertebrate predators regulate their prey? *American Naturalist* 123:125–33.

————. 1988. More thoughts on vertebrate predator regulation of prey. *American Naturalist* 132:148–54.

Errington, Paul L. 1946. Predation and vertebrate populations. *The Quarterly Review of Biology*. 21:144–77.

————. 1963. The phenomenon of predation. *American Scientist* 51:180–92.

Estes, James A. 1996. Predators and ecosystem management. *Wildlife Society Bulletin* 24:390–96.

————. 2002a. Ecological chain reactions in kelp ecosystem forests. Paper presented at *From the mountains to the sea: A conference on carnivore biology and conservation*. Monterey, California, November 18.

————. 2002b. From killer whales to kelp: Food web complexity in kelp forest ecosystems. *Wild Earth* 12(4):24–28.

————. 2004. A fitting tribute. *Conservation Biology* 18(1):279–88.

————. 2005. Carnivory and trophic connectivity in kelp forests. In *Large carnivores and the conservation of biodiversity*, ed. Justina C. Ray, Kent H. Redford, Robert S. Steneck, and Joel Berger, 61–81.

Estes, James A., Kevin Crooks, and Robert Holt. 2001. Ecological role of predators. In *Encyclopedia of Biodiversity* 4th ed. S. Levin, 857–78. San Diego: Academic Press.

Estes, J. A., E. M. Danner, D. F. Doak, B. Konar, A. M. Springer, P. D. Steinberg, M. T. Tinker, and T. M. Williams. 2004. Complex trophic interactions in kelp forest ecosystems. *Bulletin of Marine Science* 74:621–38.

Estes, James A., Douglas P. DeMaster, Daniel F. Doak, Terrie M. Williams, and Robert L. Brownell Jr., eds. 2006. *Whales, whaling, and ocean ecosystems*. Berkeley: University of California Press.

Estes, James A., and David O. Duggins. 1995. Sea otters and kelp forests in Alaska: Generality and variation in a community ecological paradigm. *Ecological Monographs* 65(1):75–100.

Estes, James A., David O. Duggins, and Galen B. Rathbun. 1989. The ecology of extinctions in kelp forest communities. *Conservation Biology* 3(3):252–64.

Estes, James A., and John F. Palmisano. 1974. Sea otters: Their role in structuring nearshore communities. *Science* 185: 1058–60.

Estes, J. A., M. L. Riedman, M. M. Staedler, M. T. Tinker, and B. E. Lyon. 2003. Individual variation in prey selection by sea otters: Patterns, causes, and implications. *Journal of Animal Ecology* 72:144–55.

Estes, James A., Norman S. Smith, and John F. Palmisano. 1978. Sea otter predation and community organization in the western Aleutian Islands, Alaska. *Ecology* 59(4):822–33.

Estes, James A., and John Terborgh. 2000. Apex predators and community dynamics: Pattern, process, and approach in some terrestrial and aquatic ecosystems. Paper presented at the Society for Conservation Biology fourteenth annual meeting, June 10, University of Montana, Missoula.

Estes, J. A., M. T. Tinker, A. M. Doroff, and D. Burn. 2005. Continuing decline of sea otter populations in the Aleutian archipelago. *Marine Mammal Science* 21:169–72.

Estes, J. A., M. T. Tinker, T. M. Williams, and D. F. Doak. 1998. Killer whale predation on sea otters linking oceanic and nearshore ecosystems. *Science* 282:473–76.

Evans, Peter G. H. 1985. Whales and dolphins. In *All the world's animals: Sea mammals*, ed. David Macdonald, 10–23.

Ewer, R. F. 1973. *The carnivores*. Ithaca, N.Y.: Cornell University Press.

Fagerstone, Kathleen A. 2002. Professional use of pesticides in wildlife management—An overview of professional wildlife damage management. *Proceedings:Vertebrate Pest Conference* 20:253–60.

Fairfax County Police Department. 2003. Fairfax Country Deer Management Report.

Fall, Michael W., and William B. Jackson. 2002. The tools and techniques of wildlife damage management—changing needs: An introduction. *International Biodeterioration and Biodegradation* 49:87–91.

Fall, Michael W., and J. Russell Mason. 2002. Developing methods for managing coyote problems—another decade of progress, 1991–2001. *Proceedings:Vertebrate Pest Conference* 20:194–200.

Feeley, Kenneth. 2003. Analysis of avian communities in Lake Guri, Venezuela, using multiple assembly rule models. *Oecologia* 137:104–113.

———. 2004. The effects of forest fragmentation and increased edge exposure on leaf litter accumulation. *Journal of Tropical Ecology* 20:709–712.

———. 2005. The role of clumped defecation in the spatial distribution of soil nutrients and the availability of nutrients for plant uptake. *Journal of Tropical Ecology* 21:99–102.

Feeley, Kenneth J., and John W. Terborgh. 2005. The effects of herbivore density on soil nutrients and tree growth in tropical forest fragments. *Ecology* 86(1):116–24.

———. 2006. Habitat fragmentation and effects of herbivore (howler monkey) abundances on bird species richness. *Ecology* 87(1):144–50.

Finke, Deborah L., and Robert F. Denno. 2004. Predator diversity dampens trophic cascades. *Nature* 429:407–10.

Fischer, Hank. 1995. *Wolf wars*. Helena, MT: Falcon Publishing.

Fish, F. E. 1998. Comparative kinematics and hydrodynamics of odontocete cetaceans: Morphological and ecological correlates with swimming performance. *Journal of Experimental Biology* 201:2867–77.

Fisher, Jonathan A. D. 2005. Exploring ecology's attic: Overlooked ideas on intertidal food webs. *Bulletin of the Ecological Society of America* 86(3):145–51.

Flader, Susan L. 1974. *Thinking like a mountain: Aldo Leopold and the evolution of an ecological attitude toward deer, wolves and forests*. Madison: University of Wisconsin Press.

Flannery, Tim. 2001. *The eternal frontier: An ecological history of North America and its peoples*. New York: Atlantic Monthly Press.

Flood, Sue. 1999. Grey killers. *BBC Wildlife*, June.

Ford, J. K. B., G. M. Ellis, D. R. Matkin, K. C. Balcomb, D. Briggs, and A. B. Morton. 2005. Killer whale attacks on minke whales. *Marine Mammal Science* 21(4):603–618.

Foreman, Dave. 2004. *Rewilding North America: A vision for conservation in the 21st century*. Washington, D.C.: Island Press.

Fortin, Daniel, Hawthorne L. Beyer, Mark S. Boyce, Douglas W. Smith, Thierry Duchesne, and Julie S. Mao. 2005. Wolves influence elk movements: Behavior shapes a trophic cascade in Yellowstone National Park. *Ecology* 86(5):1320–30.

Fortin, Daniel, Mark S. Boyce, Evelyn H. Merrill, and John M. Fryxell. 2004. Foraging costs of vigilance in large mammalian herbivores. *Oikos* 107:172–80.

Foster, Douglas. 2001. The rise of the fungus farmers. *Washington Post Magazine*, April 15: 30–40.

Foster, David R., Glenn Motzkin, Debra Bernardos, and James Cardoza. 2002. Wildlife dynamics in the changing New England landscape. *Journal of Biogeography* 29:1337–57.

Fothergill, Alastair, producer. 2001. *The blue planet: Coasts*, episode 8 of an eight-part film series by BBC/Discovery Channel.

Fowler, J., and J. Cope. 1964. Notes on harpy eagle in British Guiana. *Auk* 81:257–73.

Fox, R., S. W. Lehmkuhle, and D. H. Westendorf. 1976. Falcon visual acuity. *Science* 192:263–66.

Frank, Kenneth T., Brian Petrie, Jae S. Choi, and William C. Leggett. 2005. Trophic cascades in a cod-dominated ecosystem. *Science* 308:1621–23.

Freeman, Aaren S., and James E. Byers. 2006. Divergent induced responses to an invasive predator in marine mussel populations. *Science* 313:831–33.

Frey, Darcy. 2007. The bears among us. *New York Times Magazine*, November 25. http://www.nytimes.com/2007/11/25/magazine/25bears-t.html?ref=magazine (accessed December 21, 2007).

Frid, Alejandro, Gregory G. Baker, and Lawrence M. Dill. 2006. Do resource declines increase predation rates on Northern Pacific harbor seals? A behavior-based plausibility model. *Marine Ecology Progress Series* 312:265–75.

Fryxell, John M., John Greever, and A. R. E. Sinclair. 1988. Why are migratory ungulates so abundant? *American Naturalist* 131:781–98.

Fund for Animals. 2003. Anacapa poisoning an exercise in animal cruelty. Autumn: 15.

Gaithright, Alan. 2004. Orcas feast on harvest of gray whale calves: Killer whales team up in Monterey Bay. *San Francisco Chronicle*, May 26.

Garrott, Robert A., P. J. White, and Callie A. Vanderbilt White. 1993. Overabundance: An issue for conservation biologists? *Conservation Biology* 7(4):946–49.

Gause, G. F. 1934. *The struggle for existence*. Baltimore: Williams and Wilkins Co.

Gehrt, Stanley D. 2002. Evaluation of spotlight and road-kill surveys as indicators of local raccoon abundance. *Wildlife Society Bulletin* 30:449–56.

Gehrt, Stanley D., and William R. Clark. 2003. Raccoons, coyotes, and reflections on the mesopredator release hypothesis. *Wildlife Society Bulletin* 31(3):836–42.

Gehrt, Stanley D., George F. Huber, and Jack A. Ellis. 2002. Long-term population trends of raccoons in Illinois. *Wildlife Society Bulletin* 30:457–63.

Gehrt, Stanley D., and Suzanne Prange. 2006. Interference competition between coyotes and raccoons: A test of the mesopredator release hypothesis. *Behavioral Ecology* 18(1):204–14.

Geist, Valerius. 1998. *Deer of the world: Their evolution, behavior, and ecology*. Mechanicsburg, PA: Stackpole Books.

George, William G. 1974. Cats as predators and factors in winter shortages of raptor prey. *Wilson Bulletin* 86(4):384–96.

Gese, Eric M. 2003. Management of carnivore predation as a means to reduce livestock losses: The study of coyotes (*Canis latrans*) in North America. In *1st Workshop sobre pesquisa e conservação de carnívoros neotropicais*, Altibaia, São Paulo, Brazil, 85–102.

Gese, Eric M., and Frederick F. Knowlton. 2001. The role of predation in wildlife population dynamics. In *The role of predator control as a tool in game management*, ed. T. P. Ginnett and S. E. Henke, 7–25. San Angelo: Texas Agricultural Research and Extension Center.

Gettysburg National Military Park–Eisenhower National Historic Site. 1995. Final Environmental Impact Statement, Deer Management Plan, *Federal Register* 60, 94 (May 16): 26049.

Gittleman, John L., Stephan M. Funk, David W. Macdonald, and Robert K. Wayne, eds. 2001. *Carnivore conservation*. Cambridge: Cambridge University Press.

Gittleman, John L., and Matthew E. Gompper. 2005. Plight of predators: The importance of carnivores for understanding patterns of biodiversity and extinction risk. In *Ecology of predator-prey interactions*, ed. Pedro Barbosa and Ignacio Castellanos, 370–86. New York: Oxford University Press.

Gleeson, Scott K. 1994. Density dependence is better than ratio dependence. *Ecology* 75(6):1834–35.

Gompper, Matthew E. 2002a. The ecology of Northeast coyotes: Current knowledge and priorities for future research. *Wildlife Conservation Society Working Paper no. 17*.

———. 2002b. Top carnivores in suburbs? Ecological and conservation issues raised by colonization of northeastern North America by coyotes. *BioScience* 52(2):185–90.

Gosline, Anna. 2004. Orca attacks on gray whales in bay increase this year. *Santa Cruz Sentinel*, June 2.

Gould, Stephen J. 1989. *Wonderful life: The Burgess Shale and the nature of history*. New York: W. W. Norton and Company.

Graham, Frank Jr. 2000. The day of the condor. *Audubon*, January–February: 46–53.

Greene, Harry W. 2003. Appreciating rattlesnakes. *Wild Earth*, Summer–Fall: 28–32.

Gregg, Kathy B. 2004. Recovery of showy lady's slippers (*Cypripedium reginae* Walter) from moderate and severe herbivory by white-tailed deer (*Odocoileus virginianus* Zimmerman). *Natural Areas Journal* 24(3):232–41.

Grey, Zane. 1949. *The deer stalker*. New York: Harper.

Griggs, Jennifer A., Janet H. Rock, Christopher R. Webster, and Michael A. Jenkins. 2006. Vegetative legacy of a protected deer herd in Cades Cove, Great Smoky Mountains National Park. *Natural Areas Journal* 26:126–36.

Gude, Patricia H., Andrew J. Hansen, Ray Rasker, and Bruce Maxwell. 2006. Rates and drivers of rural residential development in the Greater Yellowstone. *Landscape and Urban Planning* 77:131–51.

Gugliotta, Guy. 2003. In Yellowstone, it's a carnivore competition. *Washington Post*, May 19.

Guinet, C. 1991. Intentional stranding apprenticeship and social play in killer whales (*Orcinus orca*). *Canadian Journal of Zoology* 69:2712–16.

Haber, Gordon. 1996. Biological, conservation, and ethical implications of exploiting and controlling wolves. *Conservation Biology* 10(4):1068–81.

Hairston, Nelson G., Frederick E. Smith, and Lawrence B. Slobodkin. 1960. Community structure, population control, and competition. *American Naturalist* 94:421–25.

Halfpenny, James C. 2003. *Yellowstone wolves in the wild*. Helena, MT: Riverbend Publishing.

Hall, E. Raymond. 1930. Predatory mammal destruction. *Journal of Mammalogy* 11(3):362–72.

Halofsky, Joshua S., and William J. Ripple. In press. Fine-scale predation risk on elk after wolf-reintroduction in Yellowstone National Park, USA. *Oecologia*.

———. In press. Linkages between wolf presence and aspen recruitment in the Gallatin elk winter range of Southwestern Montana, USA. *Forestry*.

Handwerk, Brian. 2005. Are coyotes becoming more aggressive? *National Geographic News*, June 7.

Hanscom, Greg. 1997. Politics tangles with science. *High Country News*, September 15.

Hansen, Andrew J., Richard L. Knight, John M. Marzluff, Scott Powell, Kathryn Brown, Patricia H. Gude, and Kingsford Jones. 2005. Effects of exurban development on biodiversity: Patterns, mechanisms, and research needs. *Ecological Applications* 15(6):1893–1905.

Harbo, R. 1975. Lunch with killers. *Pacific Diver* 1:21–22, 43.

Harding, Elaine K., Daniel F. Doak, and Joy D. Albertson. 2001. Evaluating the effectiveness of predator control: The non-native red fox as a case study. *Conservation Biology* 15(4):1114–22.

Hardy, Alister. 1968. Foreword to Charles Elton's Influence in Ecology. *Journal of Animal Ecology* 37(1):3–8.

Harley, Christopher D.G. 2003. Species importance and context: Spatial and temporal variation in species interactions. In *The importance of species: Perspectives on expendability and triage*, ed. Peter Kareiva and Simon A. Levin, 44–68.

Harrison, George H. 1980. *Roger Tory Peterson's dozen birding hot spots*. New York: Simon & Schuster.

Hart, Donna, and Robert W. Sussman. 2005. *Man the hunted: Primates, predators, and human evolution*. New York: Westview Press.

Haskell, David G., Jonathan P. Evans, and Neil W. Pelkey. 2006. Depauperate avifauna in plantations compared to forests and exurban areas. *Public Library of Science* 1(1):e63.

Hatfield, Brian B., Dennis Marks, M. Tim Tinker, Kellie Nolan, and Joshua Pierce. 1998. Attacks on sea otters by killer whales. *Marine Mammal Science* 14(4):888–94.

Hawes-Davis, Doug. 2000. *Killing coyote*. A film by High Plains Films.

Healy, William H. 1997. Influence of deer on the structure and composition of oak forests in central Massachusetts. In *The science of overabundance: Deer ecology and population management*, ed. William J. McShea, H. Brian Underwood, and John H. Rappole, 249–66.

Hebblewhite, Mark, Clifford A. White, Clifford G. Nietvelt, John A. McKenzie, Tomas E. Hurd, John M. Fryxell, Suzanne E. Bayley, and Paul C. Paquet. 2005. Human activity mediates a trophic cascade caused by wolves. *Ecology* 86(8):2135–44.

Hedlund, James H., Paul D. Curtis, Gwen Curtis, and Allan F. Williams. 2003. *Methods to reduce traffic crashes involving deer: What works and what does not*. Arlington, VA: Insurance Institute for Highway Safety.

Heise, K., L. G. Barrett-Lennard, E. Saulitis, C. G. Matkin and D. Bain. 2003. Examining the evidence for killer whale predation on Steller sea lions in British Columbia and Alaska. *Aquatic Mammals* 29:325–34.

Heithaus, M. R. 2005. Habitat use and group size of pied cormorants (*Phalacrocorax varius*) in a seagrass ecosystem: Possible effects of food abundance and predation risk. *Marine Biology* 147(1):27–35.

Heithaus, Michael R., and Lawrence M. Dill. 2002. Food availability and tiger shark predation risk influence bottlenose dolphin habitat use. *Ecology* 83(2):480–91.

———. 2006. Does tiger shark predation risk influence foraging habitat use by bottlenose dolphins at multiple spatial scales? *Oikos* 114:257–64.

Heithaus, M. R., L. M. Dill, G. J. Marshall, and B. Buhleier. 2002. Habitat use and foraging behavior of tiger sharks (*Galeocerdo cuvier*) in a seagrass ecosystem. *Marine Biology* 140:237–48.

Heithaus, M. R., A. Frid, and L. M. Dill. 2002. Shark-inflicted injury frequencies, escape ability, and habitat use of green and loggerhead turtles. *Marine Biology* 140:229–36.

Heithaus, Michael R., Alejandro Frid, Aaron J. Wirsing, Lars Bejder, and Lawrence M. Dill. 2005. Biology of green and loggerhead turtles under risk from tiger sharks at a foraging ground. *Marine Ecology Progress Series* 288:285–94.

Heithaus, Michael R., Alejandro Frid, Aaron J. Wirsing, Lawrence M. Dill, James W. Fourqurean, Derek Burkholder, Jordan Thompson, and Lars Bejder. 2007. State-dependent risk-taking by green sea turtles mediates top-down effects of tiger shark intimidation in a marine ecosystem. *Journal of Animal Ecology* 76(5):837–44.

Hemphill, Stephanie. 2004. Deer: The new eco-threat. Minnesota Public Radio, July 7.

Henderson, W. C. 1930. The control of the coyote. *Journal of Mammalogy* 11(3):336–53.

Henke, Scott E., and Fred C. Bryant. 1999. Effects of coyote removal on the faunal community in Western Texas. *Journal of Wildlife Management* 4:1066–81.

Herre, Wolf, and Manfred Rohrs. 1989. Domestic mammals. In *Grzimek's Encyclopedia of Mammals* 5, 570–99. New York: McGraw-Hill.

Herring, Hal. 2007. Predator hunters for the environment. *High Country News*, June 25: 10–23, 25.

Highfield, Roger. 2005. "Rewilding" could mean lions at large in US. http://www.telegraph .co.uk (accessed November 1, 2005).

Hodkinson, Ian D., and Stephen K. Coulson. 2004. Are high Arctic terrestrial food chains really that simple? The Bear Island food web revisited. *Oikos* 106(2):427–31.

Hoelzel, A. R. 1991. Killer whale predation of marine mammals at Punta Norte, Argentina: Food sharing, provisioning and foraging strategy. *Behavioral Ecology and Sociobiology* 29:197–204.

Hollenbeck, Jeff P. 2006. Multi-scale relationships between aspen and birds in the Northern Yellowstone Ecosystem. Ph.D. dissertation, Oregon State University.

Hollenbeck, Jeff P., and William J. Ripple. 2007a. Aspen and conifer heterogeneity effects on bird diversity in the northern Yellowstone ecosystem. *Western North American Naturalist* 67:92–101.

———. 2007b. Aspen patch and migratory bird relationships in the northern Yellowstone ecosystem. *Landscape Ecology* 22:1411–25.

———. In press. Aspen snag dynamics, cavity-nesting birds, and trophic cascades in Yellowstone's northern range. *Forest Ecology and Management*.

Holmes, Richard T., Thomas W. Sherry, and Franklin W. Sturges. 1986. Bird community dynamics in a temperate deciduous forest: Long-term trends at Hubbard Brook. *Ecological Monographs* 56(3):201–220.

Hooper, Robert G., Hewlette S. Crawford, and Richard F. Harlow. 1973. Bird density and diversity as related to vegetation in forest recreational areas. *Journal of Forestry* 25:766–69.

Hopper, Dale. 1998. Deer-related accidents rise sharply this month. *Fairfax Journal*, November 11.

Horner, John R., and Don Lessem. 1993. *The complete T. Rex.* New York: Simon & Schuster.

Horsley, Stephen B., Susan L. Stout, and David S. deCalesta. 2003. White-tailed deer impact on the vegetation dynamics of a northern hardwood forest. *Ecological Applications* 13(1):98–118.

Howell, A. Brazier. 1930. At the cross-roads. *Journal of Mammalogy* 11(3):377–89.

Huff, Dan E., and John D. Varley. 1999. Natural regulation in Yellowstone National Park's northern range. *Ecological Applications* 9(1):17–29.

Hunter, Mark D., and Peter W. Price. 1992. Playing chutes and ladders: Heterogeneity and the relative roles of bottom-up and top-down forces in natural communities. Ecology 73(3):724–32.

Hutchinson, G. E. 1951. Copepodology for the ornithologist. Ecology 32(3):571–77.

———. 1959. Homage to Santa Rosalia; or, why are there so many kinds of animals? American Naturalist 93(870):149–59.

Ickes, Kalen, Saara J. DeWalt, and S. Appanah. 2001. Effects of native pigs (Sus scrofa) on woody understorey vegetation in a Malaysian lowland rain forest. Journal of Tropical Ecology 17(2):191–206.

Ickes, Kalen, Christopher J. Paciorek, and Sean C. Thomas. 2005. Impacts of nest construction by native pigs (Sus scrofa) on lowland Malaysian rain forest saplings. Ecology 86(6):1540–47.

Idaho Department of Fish and Game. 2006. Effects of wolf predation on north central Idaho elk populations.

International Union for the Conservation of Nature. 1990. Predation and predator control. http://www.canids.org/1990CAP/11pcontrl.htm (accessed May 21, 2006).

———. 2007. Red List of Threatened Species. http://www.iucnredlist.org (accessed November 28, 2007).

Israelson, Brent. 2002. Wayward wolf nabbed in Utah. High Country News, December 23.

Jablonski, David. 2004. Extinction: Past and present. Nature 427:589.

Jackson, Jeremy. 2004. Habitat destruction and ecological extinction of marine invertebrates. Speech presented at the Center for Biodiversity Conservation symposium, American Museum of Natural History, New York, March 25. http://symposia.cbc.amnh.org/archives/expandingthearc/speakers/transcripts/jackson-text.html (accessed December 21, 2007).

Jackson, Jeremy B. C., Michael X. Kirby, Wolfgang H. Berger, Karen A. Bjorndal, Louis W. Botsford, Bruce J. Bourque, Roger H. Bradbury, Richard Cooke, Jon Erlandson, James A. Estes, Terence P. Hughes, Susan Kidwell, Carina B. Lange, Hunter S. Lenihan, John M. Pandolfi, Charles H. Peterson, Robert S. Steneck, Mia J. Tegner, and Robert R. Warner. 2001. Historical overfishing and the recent collapse of coastal ecosystems. Science 293:629–38.

Jaeger, Michael M., Karen M. Blejwas, Benjamin N. Sacks, Jennifer C. C. Neale, Mary M. Conner, and Dale R. McCullough. 2001. Targeting alphas can make coyote control more effective and socially acceptable. California Agriculture 55(6):32–36.

Janis, Christine. 1994. The sabertooth's repeat performances. Natural History (4):78–83.

Janzen, Daniel H., and Paul S. Martin. 1982. Neotropical anachronisms: The fruits the gomphotheres ate. Science 215:19–27.

Jaroff, Leon. 1989. Attack of the killer cats. Time, July 31.

Jefferson, Thomas A., Pam J. Stacey, and Robin W. Baird. 1991. A review of killer whale interactions with other marine mammals: Predation and co-existence. Mammal Review 21(4):151–80.

Jenkins, F. A. Jr., and S. M. Camazine. 1977. Hip structure and locomotion in ambulatory and cursorial carnivores. *Journal of Zoology* 181:351–70.

Jentz, Kathy. 2006. Doh! A deer! *Washington Examiner*, April 13: 23–24.

Johnson, Steven. 2003. Emotions and the brain: Fear. *Discover*, March: 32–39.

Johnson, Warren E., Eduardo Eizirik, and Gina M. Lento. 2001. The control, exploitation, and conservation of carnivores. In *Carnivore conservation*, ed. John L. Gittleman, Stephan M. Funk, David W. Macdonald, and Robert K. Wayne, 196–219.

Jones, Stephen B. 1993. Whitetails are changing our woodlands: Increasing white-tailed deer population may cause imbalance in forest ecosystem. *American Forests*. http://findarticles .com/p/articles/mi_m1016/is_n11-12_v99/ai_14795507 (accessed December 21, 2007).

Jontz, Sandra. 1999. Fatal crash points to deer danger. *Fairfax Journal*, October 25, A1.

Kaiser, Jocelyn. 1998. Sea otter declines blamed on hungry killers. *Science* 282:390.

Kaltenborn, Bjørn P., and Tore Bjerke. 2002. The relationship of general life values to attitudes toward carnivores. *Human Ecology Review* 9(1):55–61.

Kareiva, Peter, and Simon A. Levin, eds. 2003. *The importance of species: Perspectives on expendability and triage*. Princeton, NJ: Princeton University Press.

Kareiva, Peter, Christopher Yuan-Farrell, and Casey O'Connor. 2006. Whales are big and it matters. In *Whales, whaling, and ocean ecosystems*, ed. James A. Estes, Douglas P. Demaster, Daniel F. Doak, Terrie M. Williams, and Robert L. Brownell Jr., 379–87.

Kassen, Rees, Angus Buckling, Graham Bell, and Paul B. Rainey. 2000. Diversity peaks at intermediate productivity in a laboratory microcosm. *Nature* 406:508–512.

Kats, Lee B., and Lawrence M. Dill. 1998. The scent of death: Chemosensory assessment of predation risk by prey animals. *Ecoscience* 5(3):361–94.

Kauffman, Jason. 2007. Sheep rancher seeks peace with wolves. *Idaho Mountain Express*. http:// www.mtexpress.com/story_printer.php?ID=2005116488 (accessed September 14, 2007).

Kauffman, Matthew J., Nathan Varley, Douglas W. Smith, Daniel R. Stahler, Daniel R. MacNulty, and Mark S. Boyce. 2007. Landscape heterogeneity shapes predation in a newly restored predator-prey system. *Ecology Letters* 10(8):690–700.

Kay, Charles E. 1997. Viewpoint: Ungulate herbivory, willows, and political ecology in Yellowstone. *Journal of Range Management* 50:139–45.

———. 1998. Are ecosystems structured from the top-down or bottom-up: A new look at an old debate. *Wildlife Society Bulletin* 26:484–98.

———. 2002. Afterword: False gods, ecological myths, and biological reality. In *Wilderness and political ecology: Aboriginal influences and the original state of nature*, ed. Charles E. Kay and Randy T. Simmons, 238–61. Salt Lake City: University of Utah Press.

Kays, Roland W., and Don E. Wilson. 2002. *Mammals of North America*. Princeton, NJ: Princeton University Press.

Keigley, Richard B. 2000. Elk, beaver, and the persistence of willows in national parks: Comment on Singer et al. (1998). *Wildlife Society Bulletin* 28:448–50.

Keigley, Richard B., and Frederic H. Wagner. 1998. What is "natural"? Yellowstone elk population—a case study. *Integrative Biology* 1:133–48.

Kellert, Stephen R. 1997. *Kinship to mastery: Biophilia in human evolution and development.* Washington, D.C.: Island Press.

Kellert, Stephen R., Matthew Black, Colleen Reid Rush, and Alistair J. Bath. 1996. Human culture and large carnivore conservation in North America. *Conservation Biology* 10(4):977–90.

Kenyon Karl W. 1971. Return of the sea otter. *National Geographic* 140(4): 520–39.

———. 1975. *The sea otter in the eastern Pacific Ocean.* New York: Dover Publications.

Kidd, N. Z. C., and G. B. Lewis. 1987. Can vertebrate predators regulate their prey? A reply. *American Naturalist* 130:448–53.

Kie, John G. 2006. Review of wolf predation on north central Idaho elk populations. Unpublished report, January 9.

Kimberlin, Joanne. 2007. The problem of drivers hitting deer is getting worse. *Virginian-Pilot,* January 1.

Kirk, Ruth, and Richard D. Daugherty. 1974. *Hunters of the whale: An adventure in northwest coast archaeology.* New York: William Morrow and Company.

Kissui, Bernard M., and Craig Packer. 2004. Top-down regulation of a top predator: Lions in the Ngorongoro Crater. *Proceedings of the Royal Society of London B* 271:1867–74.

Kleiner, Kurt. 2005. Elephants and lions unleashed on North America? NewScientist.com, August 17.

Knight, Dennis H. 1994. *Mountains and plains: The ecology of Wyoming landscapes.* New Haven, CT: Yale University Press.

Knowlton, Frederick F., Eric M. Gese, and Michael M. Jaeger. 1999. Coyote depredation control: An interface between biology and management. *Journal of Range Management* 52:398–412.

Knox, W. Matt. 1997. Historical changes in the abundance and distribution of deer in Virginia. In *The science of overabundance: Deer ecology and population management,* ed. William J. McShea, H. Brian Underwood, and John H. Rappole, 27–36.

Kochman, S. 1992. Orcas feast on fresh moose. *Alaska Magazine,* October: 14.

Kostel, Ken. 2004. A top predator roars back. OnEarth, Summer.

Kraft, Helmut. 1989a. Domestic goats. In *Grzimek's Encyclopedia of Mammals* 5, 537–38. New York: McGraw-Hill.

———. 1989b. Domestic sheep. In *Grzimek's Encyclopedia of Mammals* 5, 549–50. New York: McGraw-Hill.

Krantz, Grover S. 1968. Brain size and hunting ability in earliest man. *Current Anthropology* 9(5):450–51.

Krebs, Charles J., Rudy Boonstra, Stan Boutin, and A. R. E. Sinclair. 2001. What drives the 10-year cycle of snowshoe hares? *BioScience* 51(1):25–35.

Kruuk, Hans. 2002. *Hunter and hunted: Relationships between carnivores and people.* Cambridge: Cambridge University Press.

Kuhn, Thomas S. 1970. *The structure of scientific revolutions.* Chicago: University of Chicago Press.

Kurtén, Björn. 1971. *The age of mammals.* New York: Columbia University Press.

———. 1988. *Before the Indians.* New York: Columbia University Press.

Laliberte, Andrea S., and William J. Ripple. 2003. Wildlife encounters by Lewis and Clark: A spatial analysis between Native Americans and wildlife. *BioScience* 53:1006–15.

———. 2004. Range contractions of North American carnivores and ungulates. *BioScience* 54:123–38.

Lambert, David. 1985. *The field guide to prehistoric life.* New York: Facts on File.

———. 1987. *The field guide to early man.* New York: Facts on File.

Landis, Bob, and Kathryn Pasternak, producers. 2005. *Thunderbeast.* Trailwood Films and Media.

———. 2007. *Wolf pack.* Trailwood Films and Media.

Larsen, Eric J., and William J. Ripple. 2003. Aspen age structure in the northern Yellowstone ecosystem, USA. *Forest Ecology and Management* 179:469–82.

Laundré, John W., Lucina Hernández, and Kelly B. Altendorf. 2001. Wolves, elk, and bison: Reestablishing the "landscape of fear" in Yellowstone National Park, USA. *Canadian Journal of Zoology* 79:1401–9.

Leahy, Stephen. 2007. Environment-Africa: Game parks offering protection in name only? *Inter Press Service News Agency,* September 13.

Le Bouef, Burney J. 2004. Hunting and migratory movements of white sharks in the eastern North Pacific. *Memoirs of National Institute of Polar Research,* Special Issue 58:89–100.

Leigh, Egbert Giles Jr. 2002. *A magic web: The forest of Barro Colorado Island.* New York: Oxford University Press.

———. 2003. Social conflict, biological ignorance, and trying to agree which species are expendable. In *The importance of species: Perspectives on expendability and triage,* ed. Peter Kareiva and Simon A. Levin, 239–59.

Leigh, Egbert Giles Jr., and Gerrat Jacobus Vermeij. 2002. Does natural selection organize ecosystems for the maintenance of high productivity and diversity? *Philosophical Transactions of the Royal Society of London B* 357:709–18.

Leigh, Egbert Giles Jr., S. Joseph Wright, Edward Allen Herre, and Francis E. Putz. 1993. The decline of tree diversity on newly isolated tropical islands: A test of a null hypothesis and some implications. *Evolutionary Ecology* 7(1):76–102.

Leimgruber, Peter, William J. McShea, and John H. Rappole. 1994. Predation on artificial nests in large forest blocks. *Journal of Wildlife Management* 58(2):254–60.

Leopold, Aldo. 1933. *Game Management.* New York: Charles Scribner's Sons.

———. 1936. Deer and Dauerwald in Germany, II: Ecology and policy. *Journal of Forestry* 34(5):460–66.

———. 1943. Deer irruptions. *Transactions of the Wisconsin Academy of Sciences, Arts, and Letters* 35:351–66.

———. 1944. Review of The wolves of North America. Journal of Forestry 42:928–29.

———. 1949. A Sand County Almanac, and Sketches Here and There. New York: Oxford University Press.

Leopold, Aldo, Lyle K. Sowls, David L. Spencer. 1947. A survey of over-populated deer ranges in the United States. Journal of Wildlife Management 11(2):162–77.

Lepczyk, Christopher A., Angela G. Mertig, and Jianguo Liu. 2003. Landowners and cat predation across rural-to-urban landscapes. Biological Conservation 115:191–201.

Levy, Sharon. 2006. A plague of deer. BioScience 56(9):718–21.

Lewin, Roger. 1982. Thread of life: The Smithsonian looks at evolution. Washington, D.C.: Smithsonian Books.

Lewis, Thomas A. 1987. How did the giants die? International Wildlife 17(5):5–10.

Liebenberg, Louis. 2006. Persistence hunting by modern hunter-gatherers. Current Anthropology 47(6):1017–25.

Lima, Steven L. 1998. Nonlethal effects in the ecology of predator-prey interactions. Bio-Science 48(1):25–34

———. 2002. Putting predators back into behavioral predator–prey interactions. Trends in Ecology and Evolution 17:70–75.

Lima, Steven, and Lawrence M. Dill. 1990. Behavioral decisions made under the risk of pre-dation: A review and prospectus. Canadian Journal of Zoology 68:619–40.

Lin, Sara. 2005. Sea farmers struggle to save kelp from predatory urchins. Los Angeles Times, October 14.

Lindeman, Raymond L. 1942. The trophic-dynamic aspect of ecology. Ecology 23(4):399–417.

Linden, Eugene. 1992. Search for the wolf. Time, November 9: 66–67.

Lindstedt, Stan L., James F. Hokanson, Dominic J. Wells, Steven D. Swain, Hans Hoppeler, and Vilma Navarro. 1991. Running energetics in the pronghorn antelope. Nature 353:748–50.

Lindström, Jan, Esa Ranta, Hanna Kokko, Per Lundberg, and Veijo Kaitala. 2001. From arctic lemmings to adaptive dynamics: Charles Elton's legacy in population ecology. Biological Reviews 76:129–58.

Line, Les. 1993. Silence of the songbirds. National Geographic, June: 68–90.

Litvaitis, John A., and Rafael Villafuerte. 1996. Intraguild predation, mesopredator release, and prey stability. Conservation Biology 10(2):676–77.

Logan, Kenneth A., and Linda L. Sweanor. 2001. Desert puma: Evolutionary ecology and conservation of an enduring carnivore. Washington, D.C.: Island Press.

LoGiudice, Kathleen, Richard S. Ostfeld, Kenneth A. Schmidt, and Felicia Keesing. 2003. The ecology of infectious disease: Effects of host diversity and community composition on Lyme disease risk. Proceedings of the National Academy of Sciences 100(2):567–71.

Loiselle, Bette A., and William G. Hoppes. 1983. Nest predation in insular and mainland low-land rainforest in Panama. Condor 85:93–95.

Lopez, Barry Holstun. 1978. Of wolves and men. New York: Charles Scribner's Sons.

Lopez, Gabriella Orihuela, John Terborgh, and Natalia Ceballos. 2005. Food selection by a hyperdense population of red howler monkeys (*Alouatta seniculus*). *Journal of Tropical Ecology* 21:445–50.

Lopez, Lawrence, and John Terborgh. 2007. The roles of seed predation and herbivory in the failure of tree saplings to recruit on predator-free forested islands. *Journal of Tropical Ecology* 23:129–37.

Lovell, Jeremy. 2007. Scientists add shark species to endangered list. Reuters, February 22.

Lovett, Richard A. 2006. "Killer" raccoons in Washington may be getting bum rap. *National Geographic News*, August 31.

Luoma, John R. 1997. Catfight. *Audubon*, July–August: 85–91.

Lutz, H. J. 1930. The vegetation of Heart's Content, a virgin forest in northwestern Pennsylvania. *Ecology* 11(1):1–29.

Lyons, S. Kathleen, Felisa A. Smith, and James H. Brown. 2004. Of mice, mastodons and men: Human-mediated extinctions on four continents. *Evolutionary Ecology Research* 6:339–58.

MacArthur, Robert H., and Edward O. Wilson. 1963. An equilibrium theory of insular zoogeography. *Evolution* 17(4):373–87.

———. 1967. *The theory of island biogeography*. Princeton, NJ: Princeton University Press.

Macdonald, David, ed. 1985. *All the world's animals: Sea mammals*. New York: Torstar Books.

Macdonald, David W. 2001. Postscript—carnivore conservation: Science, compromise and tough choices. In *Carnivore Conservation*, ed. John L. Gittleman, Stephan M. Funk, David W. Macdonald, and Robert K. Wayne, 524–38.

Macfadyen, Amyan. 1992. Obituary: Charles Sutherland Elton. *Journal of Animal Ecology* 61(2):499–502.

MacNulty, Daniel, Nathan Varley, and Douglas W. Smith. 2001. Grizzly bear, *Ursus arctos*, usurps bison, *Bison bison*, captured by wolves, *Canis lupus*, in Yellowstone National Park, Wyoming. *Canadian Field Naturalist* 115:495–98.

MacPhee, R. D. E., and Clare Flemming. 1999. Requiem Æternam: The last five hundred years of mammalian species extinctions. In *Extinctions in near time: Causes, contexts, and consequences*, ed. R. D. E. MacPhee, 333–71. New York: Kluwer Academic/Plenum Publishers.

Maehr, David S. 2001. What follows the elk? Restoring the large mammal fauna in the East. *Wild Earth* 11(1):50–53.

Mao, Julie S., Mark S. Boyce, Douglas W. Smith, Francis J. Singer, David J. Vales, John M. Vore, and Evelyn H. Merrill. 2005. Habitat selection by elk before and after wolf reintroduction in Yellowstone national park. *Journal of Wildlife Management* 69(4):1691–1707.

Marren, Peter. 2005. The wolf at your door. *Independent Online Edition*, August 23.

Marshall, Larry G. 1994. The terror birds of South America. *Scientific American*, February: 90–95.

Martin, Jean-Louis. 2006. Could deer overabundance impact terrestrial mollusks?—A response to Örstan. *Tentacle* 14 (January): 21–22.

Martin, Larry D. 1980. Functional morphology and the evolution of cats. *Transactions of the Nebraska Academy of Sciences* 8:141–54.

Martin, Paul S. 1967. Pleistocene overkill. *Natural History*, December: 32–38.

———. 1969. Wanted: A suitable herbivore. *Natural History*, February: 35–39.

———. 1970. Pleistocene niches for alien animals. *BioScience* 20:218–21.

———. 1973. The discovery of America. *Science* 179:969–74.

———. 1975. Sloth droppings. *Natural History*, August–September: 75–81.

———. 1987. Clovisia the beautiful! *Natural History* 96:10–13.

———. 1990. Forty thousand years of extinctions on the "planet of doom." *Paleogeography, Paleoclimatology, Paleoecology* 82:187–201.

———. 1992. The last entire earth. *Wild Earth*, Winter: 29–32.

———. 1995. Rediscovering the Desert Lab. In *Late Quaternary environments and deep history: A tribute to Paul S. Martin*, ed. David W. Steadman and Jim I. Mead, 1–24. Hot Springs, SD: Mammoth Site of Hot Springs, South Dakota, Scientific Papers 3.

———. 2005. *Twilight of the mammoths: Ice age extinctions and the rewilding of America*. Berkeley: University of California Press.

Martin, Paul S., and David A. Burney. 1999. Bring back the elephants! *Wild Earth*, Spring: 57–64.

Martin, Paul S., and Richard G. Klein, eds. 1984. *Quaternary extinctions: A prehistoric revolution*. Tucson: University of Arizona Press.

Martin, Paul S., and Christine R. Szuter. 1999. War zones and game sinks in Lewis and Clark's west. *Conservation Biology* 13(1):36–45.

Martin, Paul S., Robert S. Thompson, and Austin Long. 1985. Shasta ground sloth extinction: A test of the blitzkrieg model. In *Environments and extinctions: Man in late glacial North America*, ed. Jim I. Mead and David Meltzer, 5–14.

Mason, J. Russell, William C. Pitt, and Michael J. Bodenchuk. 2002. Factors influencing the efficiency of fixed-wing aerial hunting for coyotes in the western United States. *International Biodeterioration and Biodegradation* 49:189–97.

Matthiessen, Peter. 1987. *Wildlife in America*. Harrisonburg, VA: R. R. Donnelley and Sons.

———. 1995. *The tree where man was born*. New York: Penguin Books.

Mattson, David J., Stephen Herrero, R. Gerald Wright, and Craig M. Pease. 1996. Science and management of Rocky Mountain grizzly bears. *Conservation Biology* 16(4):1013–25.

Mattson, David J., and Troy Merrill. 2002. Extirpations of grizzly bears in the contiguous United States, 1850–2000. *Conservation Biology* 16(4):1123–36.

Mayall, Hillary. 2001. Climate change caused extinction of big ice age mammals, scientist says. *National Geographic News*, November 12. http://news.nationalgeographic.com/news/2001/11/1112_overkill.html (accessed December 21, 2007).

McCabe, Richard E., Bart W. O'Gara, and Henry M. Reeves. 2004. *Prairie ghost: Pronghorn and human interaction in early America*. Boulder: University Press of Colorado.

McCabe, Thomas R., and Richard E. McCabe. 1997. Recounting whitetails past. In *The science of overabundance: Deer ecology and population management*, ed. William J. McShea, H. Brian Underwood, and John H. Rappole, 11–26.

McCullough, Dale R. 1998. Of paradigms and philosophies: Aldo Leopold and the search for a sustainable future. Unpublished.

McDougal, Charles. 1991. Man-eaters. In *Great cats: Majestic creatures of the wild*, ed. John A. Seidensticker and Susan Lumpkin, 204–11.

McGowan, Christopher. 1997. *The raptor and the lamb: Predators and prey in the living world*. London: Allen Lane, The Penguin Press.

McGraw, W. Scott, Catherine Cooke, and Susanne Shultz. 2006. Primate remains from African crowned eagle (*Stephanoaetus coronatus*) nests in Ivory Coast's Tai Forest: Implications for primate predation and early hominid taphonomy in South Africa. *American Journal of Physical Anthropology* 131:151–65.

McIntyre, Rick, ed. 1994. *War against the wolf: America's campaign to exterminate the wolf*. Stillwater, MN: Voyageur Press.

McKinney, Michael L., and Julie L. Lockwood. 1999. Biotic homogenization: A few winners replacing many losers in the next mass extinction. *Trends in Ecology and Evolution* 14(11):450–53.

McLaren, B. E., and R. O. Peterson. 1994. Wolves, moose, and tree rings on Isle Royale. *Science* 266:1555–58.

McNamee, Thomas. 1986. *The grizzly bear*. New York: McGraw Hill.

———. 1997. *The return of the wolf to Yellowstone*. New York: Henry Holt and Company.

McNeeley, Jeffrey A. 1991. Do wild cats have a future? In *Great cats: Majestic creatures of the wild*, ed. John A. Seidensticker and Susan Lumpkin, 222–31.

McShea, William J., and John H. Rappole. 2000. Managing the abundance and diversity of breeding bird populations through manipulation of deer populations. *Conservation Biology* 14(4):1161–70.

McShea, William J., H. Brian Underwood, and John H. Rappole, eds. 1997. *The science of overabundance: Deer ecology and population management*. Washington, D.C.: Smithsonian Institution Press.

Mech, L. David. 1966. *The wolves of Isle Royale*. Washington, D.C.: U.S. Government Printing Office.

———. 1970. *The wolf: The ecology and behavior of an endangered species*. Garden City, NY: Natural History Press.

Mech, L. David, and Luigi Boitani. 2003. *Wolves: Behavior, ecology, and conservation*. Chicago and London: University of Chicago Press.

Meine, Curt. 1988. *Aldo Leopold: His life and work*. Madison: University of Wisconsin Press.

Mellink, Eric. 1995. Use of Sonoran rangelands: Lessons from the Pleistocene. In *Late Quaternary environments and deep history: A tribute to Paul S. Martin*, ed. David W. Steadman and Jim I. Mead, 50–60. Hot Springs, SD: Mammoth Site of Hot Springs, South Dakota, Scientific Papers 3.

Menge, Bruce A. 1992. Community regulation: Under what conditions are bottom-up factors important on rocky shores. *Ecology* 73(3):755–65.

———. 2003. The overriding importance of environmental context in determining the outcome of species-deletion experiments. In *The importance of species: Perspectives on expendability and triage*, ed. Peter Kareiva and Simon A. Levin, 16–43.

Menge, Bruce A., and John P. Sutherland. 1976. Species diversity gradients: Synthesis of the roles of predation, competition, and temporal heterogeneity. *American Naturalist* 110:351–69.

Meserve, Peter L., Douglas A. Kelt, W. Bryan Milstead, and Julio R. Gutiérrez. 2003. Thirteen years of shifting top-down and bottom-up control. *BioScience* 53(7):633–46.

Mezquida, Eduardo T., Steven J. Slater, and Craig W. Benkman. 2006. Sage-grouse and indirect interactions: Potential implications of coyote control on sage-grouse populations. *Condor* 108:747–59.

Mighetto, Lisa. 1991. *Wild animals and American environmental ethics.* Tucson: University of Arizona Press.

Miller, Brian, Barb Dugelby, Dave Foreman, Carlos Martinez del Río, Reed Noss, Mike Phillips, Richard Reading, Michael E. Soulé, John Terborgh, and Louisa Wilcox. 2001. The importance of large carnivores to healthy ecosystems. *Endangered Species Update* 18:202–11.

Miller, Scott G., Susan P. Bratton, and John Hadidian. 1992. Impacts of white-tailed deer on endangered and threatened vascular plants. *Natural Areas Journal* 12(2):67–74.

Mills, L. Scott, Michael E. Soulé, and Daniel F. Doak. 1993. The keystone-species concept in ecology and conservation. *BioScience* 43:219–24.

Mills, Stephen. 2004. *Tiger.* London: Firefly Books.

Miranda, João M. D., Itiberê P. Bernardi, Kauê C. Abreu, and Fernando C. Passos. 2005. Predation on *Alouatta guariba clamitans* Cabrera (Primates, Atelidae) by *Leopardus pardalis* (Linnaeus) (Carnivora, Felidae). *Revista Brasileira de zoologia* 22(3):793–95.

Mitani, John C., William J. Sanders, Jeremiah S. Lwanga, and Tammy L. Windfelder. 2001. Predatory behavior of crowned hawk-eagles (*Stephanoaetus coronatus*) in Kibale National Park, Uganda. *Behavioral Ecology and Sociobiology* 49:187–95.

Mitchell, Brian Reid. 2004. Coyote vocal communication and its application to the selective management of problem individuals. Ph.D. dissertation, University of California, Berkeley.

Mitchell, Brian R., Michael M. Jaeger, and Reginald H. Barrett. 2004. Coyote depredation management: Current methods and research needs. *Wildlife Society Bulletin* 32(4):1209–18.

Mizroch, Sally A., and Dale W. Rice. 2006. Have North Pacific killer whales switched prey species in response to depletion of the great whale populations? *Marine Ecology Progress Series* 310:235–46.

Moffat, Mark. 1995. Leafcutters: Gardeners of the ant world. *National Geographic*, July: 98–111.

Moore, Peter D. 2006. Green and pleasant trials. *Nature* 440:745–46.

Moore, Randall P., and W. Douglas Robinson. 2004. Artificial birds nests, external validity, and bias in ecological field studies. *Ecology* 85(6):1562–67.

Mora Camilo, Rebekka Metzger, Audrey Rollo, and Ransom Myers. 2007. Experimental simulations about the effects of overexploitation and habitat fragmentation on populations facing environmental warming. *Proceedings of the Royal Society B* 274(1613):1023–28.

Morin, Peter J. 2000. Biodiversity's ups and downs. *Nature* 406:463–64.

Mortensen, Camilla. 2007. Cougar kill: Will Oregon hound cougars to death? *Eugene Weekly*, July 17.

Mowat, Farley. 1996. *Sea of slaughter: A chronicle of the destruction of animal life in the North Atlantic.* Shelburne, VT: Chapters Publishing.

Mumby, Peter J., Alastair R. Harborne, Jodene Williams, Carrie V. Kappel, Daniel R. Brumbaugh, Fiorenza Micheli, Katherine E. Holmes, Craig P. Dahlgren, Claire B. Paris, and Paul G. Blackwell. 2007. Trophic cascade facilitates coral recruitment in a marine reserve. *Proceedings of the National Academy of Sciences* 104(20):8362–67.

Murdoch, William W. 1966. Community structure, population control, and competition: A critique. *American Naturalist* 100:219–26.

———. 1994. Population regulation in theory and practice. *Ecology* 75(2):271–87.

Murie, Adolph. 1985. *The wolves of Mount McKinley.* Seattle: University of Washington Press.

Murray, John A., ed. 1992. *The great bear: Contemporary writings on the grizzly.* Anchorage: Alaska Northwest Books.

Musiani, M., and P. C. Paquet. 2004. The practices of wolf persecution, protection, and restoration in Canada and the United States. *BioScience* 54:50–60.

Myers, Ransom A., Julia K. Baum, Travis D. Shepherd, Sean P. Powers, and Charles H. Peterson. 2007. Cascading effects of the loss of apex predatory sharks from a coastal ocean. *Science* 315:1846–50.

Myers, Ransom A., and Boris Worm. 2003. Rapid worldwide depletion of predatory fish communities. *Nature* 423:280–83.

———. 2005. Extinction, survival or recovery of large predatory fishes. *Philosophical Transactions of the Royal Society of London B* 360(1453):13–20.

Nash, Roderick. 1973. *Wilderness and the American mind.* Rev. ed. New Haven, CT: Yale University Press.

National Research Council. 1997. *Wolves, bears, and their prey in Alaska: Biological and social challenges in wildlife management.* Washington, D.C.: National Academies Press.

———. 2002. *Ecological dynamics on Yellowstone's northern range.* Washington, D.C.: National Academies Press.

———. 2003. *Decline of the Steller sea lion in Alaskan waters.* Washington, D.C.: National Academies Press.

Naughton-Treves, Lisa, Rebecca Grossberg, and Adrian Treves. 2003. Paying for tolerance: Rural citizens attitudes towards wolf depredation and compensation. *Conservation Biology* 17(6):1500–11.

Nelson, Michael E., and L. David Mech. 1984. Observation of a swimming wolf killing a swimming deer. Journal of Mammalogy 65:143–44.

———. 1985. Observation of a wolf killed by a deer. Journal of Mammalogy 66:187–88.

———. 1993. Prey escaping wolves, Canis lupus, despite close proximity. Canadian Field Naturalist 107:245–46.

———. 1994. A single deer stands off three wolves. American Midland Naturalist 131:207–8.

———. 2006. A three-decade dearth of deer (Odocoileus virginianus) in a wolf (Canis lupus) dominated ecosystem. American Midland Naturalist 155:373–82.

Nelson, Richard. 1997. Heart and blood: Living with deer in America. New York: Vintage Books.

Ness, Erik. 2003. Oh, deer. Discover, March: 57–71.

Newsome, A. 1990. The control of vertebrate pests by vertebrate predators. Trends in Ecology and Evolution 5:187–91.

Newton, Ian. 1990. Human impacts on raptors. In Birds of Prey, ed. Ian Newton, 190–205. New York: Facts on File.

New York Times. 1997. Scientists report rare attack by killer whales on sperm whales. November 9.

Nicastro, N. 2001. Habitats for humanity: Effects of visual affordance on the evolution of hominid antipredator communication. Evolutionary Anthropology 10:153–57.

Nicholls, Henry. 2006. Restoring nature's backbone. PLoS Biology 4(6):0893–96.

Nilsson, Greta. 2000. Persecution and hunting. In The endangered species handbook. http://www.endangeredspecieshandbook.org/persecution.php (accessed December 21, 2007).

Nogales, Manuel, Aurelio Martín, Bernie R. Tershy, C. Josh Donlan, Dick Veitch, Néstor Puerta, Bill Wood, and Jesús Alonso. 2004. A review of feral cat eradication on islands. Conservation Biology 18(2):310–19.

Norconk, M. 1996. Remembrance of Warren Kinzey (1935–1994). American Journal of Primatology 38(3):281–83.

Noss, Reed F., Howard B. Quigley, Maurice G. Hornocker, Troy Merrill, and Paul C. Paquet. 2006. Conservation biology and carnivore conservation in the Rocky Mountains. Conservation Biology 10(1):949–63.

Novacek, Michael J. 1994. A pocketful of fossils. Natural History 103 (March): 40–43.

O'Connor, Raymond J. 1991. Fading melody: Migrant songbirds are dying off with the breakup of woodlands. Sciences, January–February: 36–41.

O'Harra, Doug. 2002. Steller sleuths: Scientists work to solve the mystery of Alaska's declining sea lions. Anchorage Daily News, September 29.

———. 2004a. Scientists watch as orcas take down gray whale calf: Team researching decline in Steller sea lions happen upon coordinated kill. Anchorage Daily News, June 9.

———. 2004b. Sundry sea life in whale's belly. Anchorage Daily News, November 22.

Oksanen, Lauri, Stephen D. Fretwell, Joseph Arruda, and Pekka Niemela. 1981. Exploitation ecosystems in gradients of primary productivity. American Naturalist 118:240–61.

Olden, Julian D., and N. LeRoy Poff. 2003. Toward a mechanistic understanding and prediction of biotic homogenization. American Naturalist 162(4):442–60.

Olden, Julian D., and Thomas P. Rooney. 2006. On defining and quantifying biotic homogenization. *Global Ecology and Biogeography* 15:113–20.

Olsen, Jack. 1974. *Slaughter the animals, poison the earth.* New York: Manor Books.

Örstan, Aydin. 2006. Could deer overabundance impact terrestrial mollusks? *Tentacle* 14 (January): 20–21.

Ortega y Gasset, José. 1986. *Meditations on hunting.* New York: Charles Scribner's Sons.

Ostfeld, Richard S., Charles D. Canham, Kelly Oggenfuss, Raymond J. Winchcombe, and Felicia Keesing. 2006. Climate, deer, rodents, and acorns as determinants of variation in Lyme-disease risk. *PLoS Biology* 4(6):1058–68.

Ostfeld, Richard S., and Robert D. Holt. 2004. Are predators good for your health? Evaluating evidence for top-down regulation of zoonotic disease reservoirs. *Frontiers in Ecology and the Environment* 2(1):13–20.

Ostfeld, Richard S., and Clive G. Jones. 1999. Peril in the understory. *Audubon*, July–August: 74–82.

Ostfeld, Richard S., and Felicia Keesing. 2000. Biodiversity and disease risk: The case of Lyme disease. *Conservation Biology* 14(3):722–28.

Owen-Smith, Norman. 1987. Pleistocene extinctions: The pivotal role of megaherbivores. *Paleobiology* 13(3):351–62.

Pace, Michael L., Jonathan J. Cole, Stephen R. Carpenter, and James F. Kitchell. 1999. Trophic cascades revealed in diverse ecosystems. *Trends in Ecology and Evolution* 14(12):483–88.

Packer, Craig, and Jean Clottes. 2000. When lions ruled France. *Natural History* 109:52–57.

Packer, Craig, Robert D. Holt, Peter J. Hudson, Kevin D. Lafferty, and Andrew P. Dobson. 2003. Keeping the herds healthy and alert: Implications of predator control for infectious disease. *Ecology Letters* 6:797–802.

Packer, Craig, Dennis Ikanda, Bernard Kissui, and Hadas Kushnir. 2005. Lion attacks on humans in Tanzania. *Nature* 436:927–28.

Packer, C., D. Scheel, and A. E. Pusey. 1990. Why lions form groups: Food is not enough. *American Naturalist* 136(1):1–19.

Paine, Robert T. 1963. Ecology of the brachiopod *Glottidia pyramidata*. *Ecological Monographs* 33(3):187–213.

———. 1965. Natural history, limiting factors and energetics of the opisthobranch *Navanax inermis*. *Ecology* 46(5):603–19.

———. 1966. Food web complexity and species diversity. *American Naturalist* 100:65–75.

———. 1968. A note on trophic complexity and community stability. *American Naturalist* 102:91–93.

———. 1995. A conversation on refining the concept of keystone species. *Conservation Biology* 9(4):962–64.

———. 2000. Phycology for the mammalogist: Marine rocky shores and mammal-dominated communities—how different are the structuring processes? *Journal of Mammalogy* 81(3):637–48.

————. 2002a. Advances in ecological understanding: By Kuhnian revolution or conceptual evolution? *Ecology* 83:1553–59.

————. 2002b. Trophic control of production in a rocky intertidal community. *Science* 296:736–39.

————. 2004. Abrupt community change on a rocky shore—biological mechanisms contributing to the potential formation of an alternative state. *Ecology Letters* 7:441–45.

Paine, R. T., and Simon A. Levin. 1981. Intertidal landscapes: Disturbance and the dynamics of pattern. *Ecological Monographs* 51(2):145–78.

Paine, Robert T., and Daniel E. Schindler. 2002. Ecological pork: Novel resources and the trophic reorganization of an ecosystem. *Proceedings of the National Academy of Sciences* 99(2):554–55.

Paine, Robert T., and Robert L. Vadas. 1969. The effects of grazing by sea urchins, *Strongylocentrotus* spp., on benthic algal populations. *Limnology and Oceanography* 14:710–19.

Paine, R. T., J. T. Wooton, and P. D. Boersma. 1990. Direct and indirect effects of peregrine falcon predation on seabird abundance. *Auk* 107:1–9.

Palmer, T. S., Lee R. Dice, and Edward A. Preble. 1928. Report on the committee on conservation of land mammals. *Journal of Mammalogy* 9(4):352–54.

Palomares, Francisco, Miguel Delibes, Pablo Ferreras, and Pilar Gaona. 1996. Mesopredator release and prey abundance: Reply to Litvaitis and Villafuerte. *Conservation Biology* 10(2):678–79.

Palomares, Francisco, Pilar Gaona, Pablo Ferreras, and Miguel Delibes. 1995. Positive effects on game species of top predators by controlling smaller predator populations: An example with lynx, mongooses, and rabbits. *Conservation Biology* 9:295–305.

Patten, Michael A., and Douglas T. Bolger. 2003. Variation in top-down control of avian reproductive success across a fragmentation gradient. *Oikos* 101:479–88.

Pauly, Daniel. 1995. Anecdotes and the shifting baseline syndrome of fisheries. *Trends in Ecology and Evolution* 10(10):430.

Pauly, Daniel, and Jay Maclean. 2003. *In a perfect ocean: The state of fisheries and ecosystems in the North Atlantic Ocean.* Washington, D.C.: Island Press.

Peetz, Angela, Marilyn A. Norconk, and Warren G. Kinzey. 1992. Predation by jaguar on howler monkeys (*Alouatta seniculus*) in Venezuela. *American Journal of Primatology* 28(3):223–28.

Peres, Carlos A. 1990. A harpy eagle successfully captures an adult male red howler monkey. *Wilson Bulletin* 102(3):560–61.

Pergams, Oliver R. W., and Patricia A. Zaradic. 2006. Is love of nature in the US becoming love of electronic media? 16-year downtrend in national park visits explained by watching movies, playing video games, Internet use, and oil prices. *Journal of Environmental Management* 80:387–93.

Perkins, Sarah E., Isabella M. Cattadori, Valentina Tagliapietra, Annapaola P. Rizzoli, and Peter J. Hudson. 2006. Localized deer absence leads to tick amplification *Ecology* 87(8):1981–86.

Peterson, Karen. 2004. Wolf comeback has Wyoming officials howling. *Stateline.org*, March 11.

Peterson, Rolf O. 1995. *The wolves of Isle Royale: A broken balance.* Minocqua, WI: Willow Creek Press.

Peterson, Rolf O., and Paolo Ciucci. 2003. The wolf as a carnivore. In *Wolves: Behavior, ecology, and conservation,* ed. L. David Mech and Luigi Boitani, 104–30.

Philadelphia Inquirer. 2005. Herd control: It's time hunters change their habits. January 29.

Phillips, Michael K., and Douglas W. Smith. 1996. *The wolves of Yellowstone.* Stillwater, MN: Voyageur Press.

Pickhard, Hallie, and Robert White. 1997. Struck deer kills driver: Herd control measures revisited. *Arlington Journal,* October 23.

Pielou, E. C. 1991. *After the ice age: The return of life to glaciated North America.* Chicago: University of Chicago Press.

Pitman, Robert L., Lisa T. Ballance, Sarah I. Mesnick, and Susan J. Chivers. 2001. Killer whale predation on sperm whales: Observation and implications. *Marine Mammal Science* 17:494–507.

Pitman, Robert L., and Susan J. Chivers. 1998. Terror in black and white. *Natural History* 107(10):26–29.

Pliny the Elder. 1991. *Natural history: A selection.* Trans. and with an introduction and notes by John F. Healey. New York: Penguin Books.

Polis, Gary A. 1999. Why are parts of the world green? Multiple factors control productivity and the distribution of biomass. *Oikos* 86:3–15.

Polis, Gary A., Anna L. W. Sears, Gary R. Huxel, Donald R. Strong, and John Maron. 2000. When is a trophic cascade a trophic cascade? *Trends in Ecology and Evolution* 15(11):473–75.

Popper, Deborah Epstein, and Frank J. Popper 1987. The Great Plains: From dust to dust. *Planning,* December: 12–18.

Porter, William F., and H. Brian Underwood. 1999. Of elephants and blind men: Deer management in the U.S. national parks. *Ecological Applications* 9(1):3–9.

Potvin, François, Pierre Beaupré, and Gaétan Laprise. 2003. The eradication of balsam fir stands by white-tailed deer on Anticosti Island, Québec: A 150-year process. *Écoscience* 10(4):487–95.

Powell, Steve. 2006a. Cat-killing raccoons on prowl in west Olympia. *Olympian,* August 21.

———. 2006b. Ferocious raccoons take on area dog. *Olympian,* September 6.

———. 2006c. Raccoons ready for their close-up: National Geographic in town filming for "Nuisance Wildlife." *Olympian,* September 24.

———. 2006d. West side raccoons captivate world. *Olympian,* August 26.

Power, Mary E. 1992. Top-down and bottom-up forces in food webs: Do plants have primacy? *Ecology* 73(3):733–46.

———. 2000. What enables trophic cascades? Commentary on Polis et al. *Trends in Ecology and Evolution* 15(11):443–44.

Power, Mary E., David Tilman, James A. Estes, Bruce A. Menge, William J. Bond, L. Scott Mills, Gretchen Daily, Juan Carlos Castilla, Jane Lubchenco, and Robert T. Paine. 1996. Challenges in the quest for keystones. *BioScience* 46(8):609–20.

Primm, Steven A. 1996. A pragmatic approach to grizzly bear conservation. *Conservation Biology* 10(4):1026–35.

Primm, Steven A., and Tim W. Clark. 1996. Making sense of the policy process for carnivore conservation. *Conservation Biology* 10(4):1036–45.

Pritchard, James A. 1999. *Preserving Yellowstone's natural conditions: Science and the perception of nature.* Lincoln: University of Nebraska Press.

Pyare, Sanjay, and Joel Berger. 2003. Beyond demography and delisting: Ecological recovery for Yellowstone's grizzly bears and wolves. *Biological Conservation* 113:63–73.

Quammen, David. 2003. *Monster of God: The man-eating predator in the jungles of history and the mind.* New York: W. W. Norton and Company.

Rao, Madhu, John Terborgh, and Percy Nuñez. 2001. Increased herbivory in forest isolates: Implications for plant community structure and composition. *Conservation Biology* 15(3):624–33.

Rasker, Raymond, and Arlin Hackman. 1996. Economic development and the conservation of large carnivores. *Conservation Biology* 10(4):991–1002.

Rasmussen, D. Irvin. 1941. Biotic communities of the Kaibab Plateau, Arizona. *Ecological Monographs* 11(3):229–75.

Raven, Peter H., and George B. Johnson. 1986. *Biology.* St. Louis: Times Mirror/Mosby College Publishing.

Ray, Justina C., Kent H. Redford, Robert S. Steneck, and Joel Berger, eds. 2005. *Large carnivores and the conservation of biodiversity.* Washington, D.C.: Island Press.

Redford, Kent H., and Pamela Shaw. 1989. The terror bird still screams. *International Wildlife,* May–June: 14–16.

Reeves, Randall R., Joel Berger, and Phillip J. Clapham. 2006. Killer whales as predators of large baleen whales and sperm whales. In *Whales, whaling, and ocean ecosystems,* ed. James A. Estes, Douglas P. Demaster, Daniel F. Doak, Terrie M. Williams, and Robert L. Brownell Jr., 174–90.

Reeves, Randall R., and Tim D. Smith. 2006. A taxonomy of world whaling: Operations and eras. In *Whales, whaling, and ocean ecosystems,* ed. James A. Estes, Douglas P. Demaster, Daniel F. Doak, Terrie M. Williams, and Robert L. Brownell Jr., 82–98.

Rettig, Neil. 1977. In quest of the snatcher. *Audubon Magazine* 79:26–49.

———. 1978. Breeding behavior of the harpy eagle (*Harpia harpyja*). *Auk* 95:629–43.

———. 1995. Remote world of the harpy eagle. *National Geographic* 187(2):40–49.

Revkin, Andrew C. 2002. Out of control, deer send ecosystem into chaos. *New York Times,* November 12.

Reynolds, J. C., and S. C. Tapper. 1996. Control of mammalian predators in game management and conservation. *Mammal Review* 26:127–56.

Rice, Xan. 2007. Herdsman fights off lion but dies after hyena attack. *The Guardian,* November 21. http://www.guardian.co.uk/kenya/story/0,,2214394,00.html (accessed December 21, 2007).

Ricklefs, Robert E. 1979. Ecology. 2nd ed. New York: Chiron Press.

Ring, Ray. 2002. Wolf at the door. High Country News, May 27.

Ripple, William J., and Robert L. Beschta. 2003. Wolf reintroduction, predation risk, and cottonwood recovery in Yellowstone National Park. Forest Ecology and Management 184:299–313.

———. 2004a. Wolves and the ecology of fear: Can predation risk structure ecosystems? BioScience 54:755–66.

———. 2004b. Wolves, elk, willows, and trophic cascades in the upper Gallatin Range of southwestern Montana, USA. Forest Ecology and Management 200:161–81.

———. 2005a. Linking wolves and plants: Aldo Leopold on trophic cascades. BioScience 55(7):613–21.

———. 2005b. Refugia from browsing as reference sites for restoration planning. Western North American Naturalist 65(2):269–73.

———. 2005c. Willow thickets protect young aspen from elk browsing after wolf reintroduction. Western North American Naturalist 65(1):118–22.

———. 2006a. Linking a cougar decline, trophic cascade, and catastrophic regime shift in Zion National Park. Biological Conservation 133:397–408.

———. 2006b. Linking wolves to willows via risk-sensitive foraging by ungulates in the northern Yellowstone ecosystem. Forest Ecology and Management 230:96–106.

———. 2006c. River channel dynamics following extirpation of wolves in Yellowstone National Park, USA. Earth Surface Processes and Landforms 31(12):1525–39.

———. 2007a. Hardwood tree decline following large carnivore loss in the Great Plains, USA. Frontiers of Ecology and Environment 5:241–46.

———. 2007b. Restoring Yellowstone's aspen with wolves. Biological Conservation 138:514–19.

———. In press. Trophic cascades involving cougar, mule deer, and black oaks in Yosemite National Park. Biological Conservation.

Ripple, William J., and Eric J. Larsen. 2000. Historic aspen recruitment, elk, and wolves in northern Yellowstone National Park, USA. Biological Conservation 95:361–70.

Ripple, William J., Eric J. Larsen, Roy A. Renkin, and Douglas W. Smith. 2001. Trophic cascades among wolves, elk, and aspen on Yellowstone National Park's northern range. Biological Conservation 102:227–34.

Roach, John. 2004. Ice age bison decline not due to hunting, study says. National Geographic News, November 30. http://news.nationalgeographic.com/news/2004/11/1130_041130_bison.html (accessed December 21, 2007).

Robbins, Jim. 2003. The real world, Yellowstone: Wolves on view all the time. New York Times, July 22.

———. 2005. Hunting habits of wolves change ecological balance in Yellowstone. New York Times, October 18.

———. 2006. Deadly disease is suspected in decline of Yellowstone wolves. New York Times, January 15.

Roberts, Richard G., Timothy F. Flannery, Linda K. Ayliffe, Hiroyuki Yoshida, Jon M. Olley, Gavin J. Prideaux, Geoff M. Laslett, Alexander Baynes, M. A. Smith, Rhys Jones, and Barton L. Smith. 2001. New ages for the last Australian megafauna: Continent-wide extinction about 46,000 years ago. *Science* 292:1888–92.

Robinson, Michael J. 2005. *Predatory bureaucracy: The extermination of wolves and the transformation of the West.* Boulder: University Press of Colorado.

Robinson, Scott K., and David S. Wilcove. 1989. Conserving tropical raptors and game birds. *Conservation Biology* 3(2):192–93.

Robinson, W. Douglas. 1999. Long-term changes in the avifauna of Barro Colorado Island, Panama, a tropical forest isolate. *Conservation Biology* 13(1):85–97.

———. 2001. Changes in abundance of birds in a neotropical forest fragment over 25 years: A review. *Animal Biodiversity and Conservation* 24(2):51–65.

Robinson, W. Douglas, Ghislain Rompré, and Tara Robinson. 2005. Videography of Panama bird nests shows snakes are principal predators. *Ornitologia Neotropical* 16:187–95.

Robinson, W. Douglas, Jennifer Nesbitt Styrsky, and Jeffrey D. Brawn. 2005. Are artificial bird nests effective surrogates for estimating predation on real bird nests? A test with tropical birds. *Auk* 122(3):843–52.

Rogers, C. M., and M. J. Caro. 1998. Song sparrows, top carnivores, and nest predation: A test of the mesopredator release hypothesis. *Oecologia* 116:227–33.

Rogers, Paul. 2007. Study casts doubt on whales' comeback. *San Jose Mercury News*, September 11.

Romme, William H., Lisa Floyd-Hanna, David D. Hanna, and Elisabeth Bartlett. 2001. Aspen's ecological role in the West. USDA Forest Service Proceedings RMRS-P-18.

Romme, William H., Monica G. Turner, Linda L. Wallace, and Jennifer S. Walker. 1995. Aspen, elk, and fire in northern Yellowstone National Park. *Ecology* 76(7):2097–2106.

Rooney, Neil, Kevin McCann, Gabriel Gellner, and John C. Moore. 2006. Structural asymmetry and the stability of diverse food webs. *Nature* 442:265–69.

Rooney, Thomas P. 1997. Escaping herbivory: Refuge effects on the morphology and shoot demography of the clonal forest herb *Maianthemum canadense*. *Journal of the Torrey Botanical Society* 124(4):280–85.

———. 2001. Deer impacts on forest ecosystems: A North American perspective. *Forestry* 74:201–8.

Rooney, Thomas P., and William J. Dress. 1997a. Patterns of plant diversity in overbrowsed primary and secondary hemlock-northern hardwood forest stands. *Journal of the Torrey Botanical Society* 124(1):43–51.

———. 1997b. Species loss over sixty-six years in the ground-layer vegetation of Heart's Content, an old-growth forest in Pennsylvania, USA. *Natural Areas Journal* 17(4):297–305.

Rooney, Thomas P., and Kevin Gross. 2003. A demographic study of deer browsing impacts on *Trillium grandiflorum*. *Plant Ecology* 168:267–77.

Rooney, Thomas P., Ronald J. McCormick, Stephen L. Solheim, and Donald M. Waller. 2000. Regional variation in recruitment of hemlock seedlings and saplings in the upper Great Lakes, USA. *Ecological Applications* 10(4):1119–32.

Rooney, Thomas P., Stephen L. Solheim, and Donald M. Waller. 2002. Factors affecting the regeneration of northern white cedar in lowland forests of the upper Great Lakes region, USA. *Forest Ecology and Management* 163:119–30.

Rooney, Thomas P., and Donald M. Waller 1998. Local and regional variation in hemlock seedling establishment in forests of the upper Great Lakes region, USA. *Forest Ecology and Management* 111:211–24.

———. 2003. Direct and indirect effects of deer in forest ecosystems. *Forest Ecology and Management* 181:165–76.

Rooney, Thomas, Donald Waller, and Shannon Wiegmann. 2001. Revisiting the Northwoods—a lesson in biotic homogenization. *Wild Earth* 11:45–49.

Rooney, Thomas P., Shannon M. Wiegmann, David A. Rogers, and Donald M. Waller. 2004. Biotic impoverishment and homogenization in unfragmented forest understory communities. *Conservation Biology* 18(3):787–98.

Rosenberg, Andrew A., W. Jeffrey Bolster, Karen E. Alexander, William B. Leavenworth, Andrew B. Cooper, and Matthew G. McKenzie. 2005. The history of ocean resources: Modeling cod biomass using historical records. *Frontiers in Ecology and Environment* 3(2):78–84.

Rosenfeld, Anne Wertheim, and Robert T. Paine. 2002. *The intertidal wilderness: A photographic journey through Pacific coast tidepools*. Berkeley: University of California Press.

Rubenstein, Dustin R., Daniel I. Rubenstein, Paul W. Sherman, and Thomas A. Gavin. 2006. Pleistocene park: Does re-wilding North America represent sound conservation for the 21st century? *Biological Conservation* 132:232–38.

Russell, F. Leland, David B. Zippin, and Norma L. Fowler. 2001. Effects of white-tailed deer (*Odocoileus virginianus*) on plants, plant populations and communities: A review. *American Midland Naturalist* 146(1):1–26.

Russell, Gareth J. 2004. Valuing species in context. *Ecology* 85(1):292–94.

Sæather, Bernt-Erik. 1999. Top dogs maintain diversity. *Nature* 400:510–11.

Sargeant, Alan B., Raymond J. Greenwood, Marsha A. Sovada, and Terry L. Shafer. 1994. Distribution and abundance of predators that affect duck production—Prairie Pothole Region. U.S. Fish and Wildlife Service, Resource Publication 194. Jamestown, ND: Northern Prairie Wildlife Research Center Online. http://www.npwrc.usgs.gov/resource/birds/predator/index.htm (accessed December 21, 2007).

Saulitis, Eva, Craig Matkin, Lance Barrett-Lennard, Kathy Heise, and Graeme Ellis. 2000. Foraging strategies of sympatric killer whale (*Orcinus orca*) populations in Prince William Sound, Alaska. *Marine Mammal Science* 16(1):94–109.

Savage, Robert J. G. 1986. *Mammal evolution: An illustrated guide*. New York: Facts on File.

Schaller, George B. 1973. *Golden shadows, flying hooves*. New York: Dell.

Schaller, George B., and Gordon R. Lowther. 1969. The relevance of carnivore behavior to the study of early hominids. *Southwest Journal of Anthropology.* 25:307–341.

Schauber, Eric M., Richard S. Ostfeld, and Andrew S. Evans Jr. 2005. What is the best predictor of annual Lyme disease incidence: Weather, mice, or acorns? *Ecological Applications* 15(2):575–86.

Scheffer, Marten, and Stephen R. Carpenter. 2003. Catastrophic regime shifts in ecosystems: Linking theory to observation. *Trends in Ecology and Evolution* 18(12):648–56.

Scheffer, Marten, Steve Carpenter, Jonathan A. Foley, Carl Folke, and Brian Walker. 2001. Catastrophic shifts in ecosystems. *Nature* 413:591–96.

Schirber, Michael. 2004. Surviving extinction: Where woolly mammoths endured. *Live Science*, October 19. http://www.livescience.com/animals/041019_Mammoth_Island.html (accessed December 21, 2007).

Schlaepfer, Martin A. 2005. Re-wilding: A bold plan that needs native megafauna. *Nature* 437:951.

Schmidt, Kenneth A. 2003. Nest predation and population declines in Illinois songbirds: A case for mesopredator effects. *Conservation Biology* 17(4):1141–50.

Schmidt, Kenneth A., and Richard S. Ostfeld. 2001. Biodiversity and the dilution effect in disease ecology. *Ecology* 82:609–19.

———. 2003. Songbird populations in fluctuating environments: Nest predator responses to pulsed resources. *Ecology* 84(2):406–15.

Schmitz, Oswald J. 2003. Top predator control of plant biodiversity and productivity in an old-field ecosystem. *Ecology Letters* 6:156–63.

———. 2004. Perturbation and abrupt shift in trophic control of biodiversity and productivity. *Ecology Letters* 7:403–9.

Schmitz, Oswald J., Andrew P. Beckerman, and Kathleen M. O'Brien. 1997. Behaviorally mediated trophic cascades: Effects of predation risk on food web interactions. *Ecology* 78(5):1388–99.

Schmitz, Oswald J., Peter A. Hambäck, and Andrew P. Beckerman. 2000. Trophic cascades in terrestrial systems: A review of the effects of top carnivore removals on plants. *American Naturalist* 155(2):141–53.

Schmitz, Oswald J., Vlastimil Krivan, and Ofer Ovadia. 2004. Trophic cascades: the primacy of trait-mediated indirect interactions. *Ecology Letters* 7:153–63.

Schoener, Thomas W., and David A. Spiller. 2003. Effects of removing a vertebrate versus an invertebrate predator on a food web, and what is their relative importance? In *The importance of species: Perspectives on expendability and triage*, ed. Peter Kareiva and Simon A. Levin, 69–84.

Schoener, T. W., D. A. Spiller, and J. B. Losos. 2001. Predators increase the risk of catastrophic extinction of prey populations. *Nature* 412:183–86.

Schullery, Paul. 1996. *The Yellowstone wolf: A guide and sourcebook.* Worland, WY: High Plains Publishing.

Schwabe, Kurt A., Peter W. Schuhmann, Michael J. Tonkovich and Ellen Wu. 2002. An analysis of deer-vehicle collisions: The case of Ohio. In *Human conflicts with wildlife: Economic considerations*, Proceedings of the Third NWRC Special Symposium, National Wildlife Research

Center, Fort Collins, CO, ed. Larry Clark, Jim Hone, John A. Shivik, Richard A. Watkins, Kurt C. VerCauteren, and Jonathan K. Yoder, 91–103.

Schwartz, Charles C., Jon E. Swenson, and Sterling D. Miller. 2003. Large carnivores, moose, and humans: A changing paradigm of predator management in the 21st century. *Alces* 39:41–63.

Schwartz, Noaki. 2006. Raccoons invade California enclave. Associated Press, November 17.

Schweiger, Larry J. 2007. Fostering a sense of wonder. *National Wildlife Magazine* 45(5). http://www.nwf.org/nationalwildlife/article.cfm?issueID=116&articleID=1500 (accessed September 12, 2007).

Seidensticker, John A., and Susan Lumpkin, eds. 1991. *Great cats: Majestic creatures of the wild*. Emmaus, PA: Rodale Press.

Seligman, Daniel M. 1998. Oh, deer—revisited: Sharpshooter plan okayed. *Mt. Vernon Gazette*, November 19.

Séquin, Eveline S., Michael M. Jaeger, Peter F. Brussard, and Reginal H. Barrett. 2003. Wariness of coyotes to camera traps relative to social status and territory boundaries. *Canadian Journal of Zoology* 81:2015–25.

Sergio, Fabrizio, Ian Newton, and Luigi Marchesi. 2005. Top predators and biodiversity. *Nature* 436:192.

Sergio, Fabrizio, Ian Newton, Luigi Marchesi, and Paolo Pedrini. 2006. Ecologically justified charisma: Preservation of top predators delivers biodiversity conservation. *Journal of Applied Ecology* 43:1049–55.

Seton, Ernest Thompson. 1926. *Wild animals I have known*. New York: Scribners.

———. 1929. *Lives of game animals*. 4 vols. Garden City, NY: Doubleday, Doran, and Co.

Shepard, Paul. 1996. *The only world we've got: A Paul Shepard reader*. San Francisco: Sierra Club Books.

———. 1998. *Coming home to the Pleistocene*. Washington, D.C.: Island Press.

Shissler, Bryon P. 2006. Testimony before the House Agriculture and Rural Affairs Committee regarding deer damage and related issues. April 4.

Shivak, John A. 2004. Non-lethal alternatives for predator management. *Sheep and Goat Research Journal* 19:64–71.

Shurin, Jonathan B., Elizabeth T. Borer, Eric W. Seabloom, Kurt Anderson, Carol A. Blanchette, Bernardo Broitman, Scott D. Cooper, and Benjamin S. Halpern. 2002. A cross-ecosystem comparison of the strength of trophic cascades. *Ecology Letters* 5:785–91.

Sieving, Kathryn E. 1992. Nest predation and differential insular extinction among selected forest birds of central Panama. *Ecology* 73(6): 2310–28.

Silliman, Brian Reed, and Mark D. Bertness. 2002. A trophic cascade regulates salt marsh primary production. *Proceedings of the National Academy of Sciences* 99(16):10500–05.

Simberloff, Daniel. 1998. Flagships, umbrellas, and keystones: Is single-species management passé in the landscape era? *Biological Conservation* 83(3):247–57.

———. 2003. Community and ecosystem impacts of single-species extinctions. In *The importance of species: Perspectives on expendability and triage*, ed. Peter Kareiva and Simon A. Levin, 221–33.

Simenstad, Charles A., James A. Estes, and Karl W. Kenyon. 1978. Aleuts, sea otters, and alternate stable state communities. Science 200:403–11.

Sinclair, A. R. E. 1995. Population consequences of predation-sensitive foraging: The Serengeti wildebeest. Ecology 76(3):882–91.

———. 1997. Carrying capacity and the overabundance of deer: A framework for management. In The science of overabundance: Deer ecology and population management, ed. William J. McShea, H. Brian Underwood, and John H. Rappole, 380–94.

Singer, Francis J., Linda C. Zeigenfuss, and David T. Barnett. 2000. Elk, beaver, and the persistence of willows in national parks: Response to Keigley. Wildlife Society Bulletin 28(2):451–53.

Slobodkin, L. B., F. E. Smith, and N. G. Hairston. 1967. Regulation in terrestrial ecosystem and the implied balance of nature. American Naturalist 101:109–24.

Smil, Vaclav. 2002. Eating meat: Evolution, patterns, and consequences. Population and Development Review 28(4):599–639.

Smith, Christopher Irwin. 2005. Re-wilding: Introductions could reduce biodiversity. Nature 437:318.

Smith, Douglas W., and Gary Ferguson. 2006. Decade of the wolf: Returning the wild to Yellowstone. Guildford, CT: Lyons Press.

Smith, Douglas W., Rolf O. Peterson, and Douglas B. Houston. 2003. Yellowstone after wolves. BioScience 53:330–40.

Smith, Douglas W., Daniel R. Stahler, Debra S. Guernsey, Matthew Metz, Abigail Nelson, Erin Albers, and Richard McIntyre. 2007. Yellowstone wolf project: Annual report, 2006. National Park Service, Yellowstone Center for Resources, Yellowstone National Park, Wyoming, YCR-2007-01.

Snape, William J. III. 2000. Big win for wolves. Defenders 75(2):6–7.

Soulé, Michael E., Douglas T. Bolger, Allison C. Alberts, John Wright, Marina Sorice, and Scott Hill. 1988. Reconstructed dynamics of rapid extinctions of chaparral-requiring birds in urban habitat islands. Conservation Biology 2(1):75–92.

Soulé, Michael E., James A. Estes, Joel Berger, and Carlos Martinez del Rio. 2003. Ecological effectiveness: Conservation goals for interactive species. Conservation Biology 17(5):1238–50.

Soulé, Michael E., James A. Estes, Brian Miller, and Douglas L. Honnold. 2005. Strongly interacting species: Conservation policy, management, and ethics. BioScience 55(2):168–76.

Soulé, Michael E., and Reed Noss. 1998. Rewilding and biodiversity: Complementary goals for continental conservation. Wild Earth 8(3):18–28.

Soulé, Michael E., and John Terborgh, eds. 1996. Continental conservation: Scientific foundations of regional reserve networks. Washington, D.C.: Island Press.

Soulé, Michael E., and Bruce A. Wilcox, eds. 1980. Conservation biology: An evolutionary-ecological perspective. Sunderland, MA: Sinauer Associates.

Southwest Fisheries Science Center. 1997. Sperm whales attacked by killer whales. News release, November 7.

Southwood, Richard, and J. R. Clarke. 1999. Charles Sutherland Elton. *Biographical Memoirs of Fellows of the Royal Society of London* 45:129–46.

Sovada, Marsha A., Alan B. Sargeant, and James W. Grier. 1995. Differential effects of coyotes and red foxes on duck nest success. *Journal of Wildlife Management* 59(1):1–8.

Spotts, Peter N. 2005. New plan for the Great Plains: Bring back the Pleistocene. *Christian Science Monitor*, August 18.

Springer, A. M., J. A. Estes, G. B. van Vliet, T. M. Williams, D. F. Doak, E. M. Danner, K. A. Forney, and B. Pfister. 2003. Sequential megafaunal collapse in the North Pacific Ocean: An ongoing legacy of industrial whaling? *Proceedings of the National Academy of Sciences* 100(21):12223–28.

Stahler, Daniel, Bernd Heinrich, and Douglas Smith. 2002. Common ravens, *Corvus corax*, preferentially associate with grey wolves, *Canis lupus*, as a foraging strategy in winter. *Animal Behaviour* 64(2):283–90.

Stallcup, Richard. 1991. Cats take a heavy toll on songbirds: A reversible catastrophe. *Point Reyes Bird Observatory Observer*, Spring–Summer: 8–9.

Stark, Mike. 2006. UM economist: Wolves a big moneymaker. *Billings Gazette*, April 7.

Steadman, David W., and Paul S. Martin. 1984. Extinction of birds in the Late Pleistocene of North America. In *Quaternary extinctions: A prehistoric revolution*, ed. Paul S. Martin and Richard G. Klein, 466–77.

Steinmetz, Steven L. Kohler, and Daniel A. Soluk. 2002. Birds are overlooked top predators in aquatic food webs. *Ecology* 84(5):1324–28.

Stejneger, Leonhard. 1887. How the great northern sea-cow (*Rytina*) became exterminated. *American Naturalist* 21(12):1047–54.

Steneck, Robert S., Michael H. Graham, Bruce J. Bourque, Debbie Corbett, Jon M. Erlandson, James A. Estes, and Mia J. Tegner. 2002. Kelp forest ecosystems: Biodiversity, stability, resilience and their future. *Environmental Conservation* 29(4):436–59.

Steneck, Robert S., and Enric Sala. 2005. Large marine carnivores: Trophic cascades and top-down controls in coastal ecosystems past and present. In *Large carnivores and the conservation of biodiversity*, ed. Justina Ray, Kent H. Redford, Robert S. Steneck, and Joel Berger, 110–37.

Stevens, William K. 1995a. Triumph and loss overlap as wolves return to Yellowstone. *New York Times*, September 12.

———. 1995b. Wolf's howl heralds change for old haunts. *New York Times*, January 31.

———. 1998. Debating nature of nature in Yellowstone. *New York Times*, June 23.

———. 1999. Search for missing sea otters turns up a few surprises. *New York Times*, January 5.

———. 2000. Lost rivets and threads, and ecosystems pulled apart. *New York Times*, July 4.

Stewart, Doug. 2006. City Slickers. *National Wildlife*, October–November. http://www.nwf.org/nationalwildlife/article.cfm?issueID=110&articleID=1404 (accessed December 18, 2007).

Stinson, Kristina A., Stuart A. Campbell, Jeff R. Powell, Benjamin E. Wolfe, Ragan M. Callaway,

Giles C. Thelen, Steven G. Hallett, Daniel Prati, and John N. Klironomos. 2006. Invasive plant suppresses the growth of native tree seedlings by disrupting belowground mutualisms. PLoS Biology 4(5):727–31.

Stockton, Stephen A., Sylvain Allombert, Anthony J. Gaston, and Jean-Louis Martin. 2005. A natural experiment on the effects of high deer densities on the native flora of coastal temperate rain forests. Biological Conservation 126:118–28.

Stolzenburg, William. 1990a. Back when life got hard: How animals developed skeletons in a primordial arms race. Washington Post, September 30.

———. 1990b. When life got hard: Animal skeletons emerged abruptly. Why the big hurry? Science News 138:120–23.

———. 1994. New views of ancient times. Nature Conservancy, September–October: 10–15.

———. 2003a. The long rangers. Nature Conservancy 53(3):34–43.

———. 2003b. Playing god in a tide pool. Nature Conservancy 54(4):42–50.

———. 2005a. Grizzlies in the midst. Nature Conservancy 55(4):80.

———. 2005b. Shifting baselines. Nature Conservancy 55(1):80.

———. 2005c. So others might live. Nature Conservancy 55(3):80.

———. 2006a. Us or them. Conservation in Practice 7(4):14–21

———. 2006b. Where the wild things were. Conservation in Practice 7(1):28–34.

Stratford, Jeffrey A., and W. Douglas Robinson. 2005. Gulliver travels to the fragmented tropics: Geographic variation in mechanisms of avian extinction. Frontiers in Ecology and Environment 3(2):91–98.

Stromayer, Karl A. K., and Robert J. Warren. 1997. Are overabundant deer herds in the eastern United States creating alternate stable states in forest plant communities? Wildlife Society Bulletin 25(2):227–34.

Strong, Donald R. 1992. Are trophic cascades all wet? Differentiation and donor-control in speciose ecosystems. Ecology 73(3):747–54.

Suominen, Otso. 2006. Cervids and gastropods—a response to Örstan. Tentacle 14 (January): 22.

Surovell, Todd, Nicole Waguespack, and P. Jeffrey Brantingham. 2005. Global archaeological evidence for proboscidean overkill. Proceedings of the National Academy of Sciences 102(17):6231–36.

Swarner, Matthew. 2004. Human-carnivore conflict over livestock: The African wild dog in central Botswana. Paper presented at the Center for African Studies Breslauer Symposium on Natural Resource Issues in Africa, University of California, Berkeley, March 5.

Tarpy, Cliff. 1979. Killer whale attack! National Geographic, April: 542–45.

Tedford, Richard H. 1994a. Caught in time. Natural History (4):90–91.

———. 1994b. Key to the carnivores. Natural History (4):74–76.

Temple, Stanley A. 1990. Conservation and management. In Birds of Prey, ed. Ian Newton, 208–25. New York: Facts on File.

Terborgh, John. 1974. Preservation of natural diversity: The problem of extinction prone species. BioScience 24:153–69.

————. 1983. Five new world primates: A study in comparative ecology. Princeton, NJ: Princeton University Press.

————. 1988. The big things that run the world—A sequel to E. O. Wilson. *Conservation Biology* 2(4):402–3.

————. 1989. *Where have all the birds gone?: Essays on the biology and conservation of birds that migrate to the American tropics.* Princeton, NJ: Princeton University Press.

————. 1992. *Diversity and the tropical rain forest.* New York: Scientific American Library.

————. 1999. *Requiem for nature.* Washington, D.C.: Island Press.

————. 2005. The green world hypothesis revisited. In *Large carnivores and the conservation of biodiversity,* ed. Justina C. Ray, Kent H. Redford, Robert S. Steneck, and Joel Berger, 82–99.

Terborgh, John, James A. Estes, Paul Paquet, Katherine Ralls, Diane Boyd-Heger, Brian J. Miller, and Reed F. Noss. 1999. The role of top carnivores in regulating terrestrial ecosystems. In *Continental conservation: Scientific foundations of regional reserve networks,* ed. Michael E. Soulé and John Terborgh, 39–64.

Terborgh, John, Kenneth Feeley, Miles Silman, Percy Nuñez, and Bradley Balukjian. 2006. Vegetation dynamics of predator-free land-bridge islands. *Journal of Ecology* 94:253–63.

Terborgh, J., and C. H. Janson. 1986. The socioecology of primate groups. *Annual Review of Ecology and Systematics* 17:111–35.

Terborgh, John, Lawrence Lopez, Percy Nuñez, Madhu Rao, Ghazala Shahabuddin, Gabriela Orihuela, Mailen Riveros, Rafael Ascanio, Greg H. Adler, Thomas D. Lambert, and Luis Balbas. 2001. Ecological meltdown in predator-free forest fragments. *Science* 294:1923–26.

Terborgh, John, Lawrence Lopez, and José Tello S. 1997. Bird communities in transition: The Lago Guri Islands. *Ecology* 78(5):1494–1501.

Terborgh, John, and Blair Winter. 1980. Some causes of extinction. In *Conservation biology: An evolutionary-ecological perspective,* ed. Michael E. Soulé and Bruce A. Wilcox, 119–133.

Ternullo, Richard, and Nancy Black. 2002. Predation behavior of transient killer whales in Monterey Bay, California. Report presented at the International Orca Symposium, Chize, France, September.

Theil, Stefan, 2006. The new jungles. *Newsweek,* July 3–10.

Thomas, Lewis. 1974. *The lives of a cell: Notes of a biology watcher.* Toronto: Bantam Books.

Thompson, Nicholas. 2003. Not just a predator: Wolves bring a surprising ecological recovery to Yellowstone. *Boston Globe,* September 30.

Timm, Robert M., Rex O. Baker, Joe R. Bennett, and Craig C. Coolahan. 2004. Coyote attacks: An increasing suburban problem. Agriculture and Natural Resources Research and Extension Centers, Hopland Research and Extension Center (University of California).

Treves, Adrian 1999. Has predation shaped the social systems of arboreal primates? *International Journal of Primatology* 20:(1)35–67.

Treves, Adrian, Randle R. Jurewicz, Lisa Naughton-Treves, Robert A. Rose, Robert C. Willging, and Adrian P. Wydeven. 2002. Wolf depredation on domestic animals in Wisconsin, 1976–2000. *Wildlife Society Bulletin* 30(1):231–41.

Treves, Adrian, and L. Naughton-Treves. 1999. Risk and opportunity for humans coexisting with large carnivores. Journal of Human Evolution 36: 275–82.

Trites, A. W., V. Christensen, and D. Pauly. 2006. Effects of fisheries on ecosystems: Just another top predator? In Top predators in marine ecosystems: Their role in monitoring and management, ed. Ian Boyd, Sarah Wanless, and C. J. Camphuysen, 11–27. Cambridge: Cambridge University Press.

Trites, Andrew W., Volker B. Deecke, Edward J. Gregr, John K. B. Ford, Peter F. Olesiuk. 2007. Killer whales, whaling, and sequential megafaunal collapse in the North Pacific: A comparative analysis of the dynamics of marine mammals in Alaska and British Columbia following commercial whaling. Marine Mammal Science 23(4):751–65.

Tu, Mandy. 2000. Element stewardship abstract for Microstegium vimineum. Nature Conservancy's Wildland Invasive Species Program.

Turchi, Gale M., Patricia L. Kennedy, Dean Urban, and Dale Hein. 1995. Bird species richness in relation to isolation of aspen habitats. Wilson Bulletin 107:463–74.

Turchin, Peter. 2003. Complex population dynamics: A theoretical/empirical synthesis. Princeton, NJ: Princeton University Press.

Turner, Pamela S. 2004. Showdown at sea. National Wildlife 42(6): http://www.nationalwildlife .org/nationalwildlife/article.cfm?articleID=991&issueID=70 (accessed December 18, 2007).

USDA Forest Service. 2004. The state of the forest: A snapshot of Pennsylvania's updated forest inventory 2004. NA-FR-03-04.

U.S. Fish and Wildlife Service. 1961. International fur seal treaty negotiated 50 years ago. Department of the Interior Information Service.

U.S. General Accounting Office. 1990. Wildlife management: Effects of animal damage control program on predators. RCED-90-149.

———. 1995. Animal damage control program: Efforts to protect livestock from predators. RCED-96-3.

———. 2001. Wildlife services program: Information on activities to manage wildlife damage. GAO-02-138.

Van Ballenberghe, Victor. 2004. Predator control in Alaska: An analysis of current predator control programs and the recommendations of the 1997 NRC report, Wolves, bears and their prey in Alaska. Paper presented at Carnivores 2004: Expanding partnerships in carnivore conservation Conference, November 16, Santa Fe, New Mexico.

Van Valkenburgh, Blaire. 1994. Tough times in the tar pits. Natural History (4):84–85.

———. 1999. Major patterns in the history of carnivorous mammals. Annual Review of Earth and Planetary Sciences 27: 463–93.

———. 2001. Predation in saber-tooth cats. In Paleobiology II, ed. Derek E. G. Briggs and Peter R. Crowther, 420–24. Oxford: Blackwell Science.

Van Valkenburgh, Blaire, and Fritz Hertel. 1993. Tough times at La Brea: Tooth breakage in large carnivores of the Late Pleistocene. Science 261:456–59.

Van Valkenburgh, Blaire, and Ralph E. Molnar. 2003. Dinosaurian and mammalian predators compared. *Paleobiology* 28:527–43.

Van Valkenburgh, Blaire, Xiaoming Wang, and John Damuth. 2004. Cope's rule, hypercarnivory, and extinction in North American canids. *Science* 306:101–4.

Vermeij, Geerat J. 1974. Marine faunal dominance and molluscan shell form. *Evolution* 28:656–64.

———. 1976. Interoceanic differences in vulnerability of shelled prey to crab predation. *Nature* 260:135–36.

———. 1977. The Mesozoic marine revolution: Evidence from snails, predators and grazers. *Paleobiology* 3:245–58.

———. 1982a. Gastropod shell form, repair, and breakage in relation to predation by the crab *Calappa*. *Malacologia* 23(1):1–12.

———. 1982b. Unsuccessful predation and evolution. *American Naturalist* 120(6):701–20.

———. 1989. The origin of skeletons. *Palaios* 4:585–89.

———. 1997. *Privileged hands: A scientific life*. New York: W. H. Freeman and Company.

Verrengia, Joseph B. 2005. Group wants to transplant African animals. http://www.usatoday.com/news/nation/2005-08-17-wild-america_x.htm (accessed August 18, 2005).

Vila, Bruno, Frédéric Guibal, and Jean-Louis Martin. 2001. Impact of browsing on forest understory in Haida Gwaii: A dendro-ecological approach. *Laskeek Bay Research* 10:62–71.

Vos, Daniel, Lori T. Quakenbush, and Barbara A. Mahoney. 2006. Documentation of sea otters and birds as prey for killer whales. *Marine Mammal Science* 21(1):201–5.

Vourc'h, Gwenaël, Jean-Louis Martin, Patrick Duncan, José Escarré and Thomas P. Clausen. 2001. Defensive adaptations of *Thuja plicata* to ungulate browsing: A comparative study between mainland and island populations. *Oecologia* 126:84–93.

Vourc'h, Gwenaël, John Russell, Dominique Gillon, and Jean-Louis Martin. 2003. Short-term effect of defoliation on terpene content of *Thuja plicata*. *Écoscience* 10(2):161–67.

Vourc'h, Gwenaël, Bruno Vila, Dominique Gillon, José Escarré, Frédéric Guibal, Hervé Fritz, Thomas P. Clausen, and Jean-Louis Martin. 2002. Disentangling the causes of damage variation by deer browsing on young *Thuja plicata*. *Oikos* 98:271–83.

Vucetich, John A., and Rolf O. Peterson. 2004. The influence of top-down, bottom-up, and abiotic factors on the moose (*Alce alces*) population of Isle Royale. *Proceedings of the Royal Society of London B* 271:183–89.

Vucetich, John A., Rolf O. Peterson, and Carrie L. Schaefer. 2002. The effect of prey and predator densities on wolf predation. *Ecology* 83(11):3003–13.

Vucetich, John A., Rolf O. Peterson, and Thomas A. Waite. 2004. Raven scavenging favours group foraging in wolves. *Animal Behaviour* 67:1117–26.

Vucetich, John A., Douglas W. Smith, and Daniel R. Stahler. 2005. Influence of harvest, climate and wolf predation on Yellowstone elk, 1961–2004. *Oikos* 111(2):259–70.

Wade, Paul R. Lance G. Barrett-Lennard, Nancy A. Black, Vladimir N. Burkanov, Alexander M. Burdin, John Calambokidis, Sal Cerchio, Marilyn E. Dahlheim, John K. B. Ford, Nancy A. Friday, Lowell W. Fritz, Jeff K. Jacobsen, Thomas R. Loughlin, Craig O. Matkin, Dena

R. Matkin, Shannon M. McCluskey, Amee V. Mehta, Sally A. Mizroch, Marcia M. Muto, Dale W. Rice, Robert J. Small, Janice M. Straley, Glenn R. Van Blaricom, and Phillip J. Clapham. 2007. Killer whales and marine mammal trends in the North Pacific—a re-examination of evidence for sequential megafauna collapse and the prey-switching hypothesis. *Marine Mammal Science* 23(4):766–802.

Wade, Paul, John W. Durban, Janice M. Waite, Alexandre N. Zerbinik, and Marilyn E. Dahlheim. 2003. Surveying killer whale abundance and distribution in the Gulf of Alaska and Aleutian Islands. *Alaska Fisheries Science Center Quarterly Report*, October–December: 1–16.

Wagner, Frederic H., and Charles E. Kay. 1991. "Natural" or "healthy" ecosystems: Are U.S. national parks providing them? In *Humans as components of ecosystems*, ed. Mark J. McDonnell and Stewart T. A. Pickett, 257–70. New York: Springer-Verlag.

Walker, Ernest P. 1983. *Mammals of the world*, 4th ed., vol. II. London: Johns Hopkins University Press.

Wallace, Alfred Russel. 1876. *The geographical distribution of animals*. Vol. 1. New York: Harper and Brothers.

Waller, Donald M., and William S. Alverson. 1997. The white-tailed deer: A keystone herbivore. *Wildlife Society Bulletin* 25(2):217–26.

Waller, Donald M., and Thomas P. Rooney. 2004. Nature is changing in more ways than one. *Trends in Ecology and Evolution* 19:6–7.

Waller, Don, and Sarah Wright. 2006. Trouble in the understory. *Woodland Management* 27(2):20–23.

Weaver, John. 1978. The wolves of Yellowstone. Washington, D.C.: USDI National Park Service, Natural Resources Report no. 14.

Weaver, John L., Paul C. Paquet, and Leonard F. Ruggiero. 1996. Resilience and conservation of large carnivores in the Rocky Mountains. *Conservation Biology* 10(4):964–76.

Weber, William, and Alan Rabinowitz. 1996. A global perspective on large carnivore conservation. *Conservation Biology* 10(4):1046–54.

Webster, Christopher R., Michael A. Jenkins, and Janet H. Rock. 2005a. Long-term response of spring flora to chronic herbivory and deer exclusion in Great Smoky Mountains National Park, USA. *Biological Conservation* 125:297–307.

———. 2005b. Twenty years of forest change in the woodlots of Cades Cove, Great Smoky Mountains National Park. *Journal of the Torrey Botanical Society* 132(2):280–92.

Weise, Elizabeth. 2005. Beasts of both worlds; scientists propose 'rewilding' USA. *USA Today*, August 18.

Wellings, H. P. 1964. Shore whaling at Twofold Bay. *Eden* 1–15.

Wells, H. G., Julian S. Huxley, and G. P. Wells. 1929. *The science of life*. New York: Literary Guild.

White, Clifford A., Michael C. Feller, and Suzanne Bayley. 2003. Predation risk and the functional response of elk-aspen herbivory. *Forest Ecology and Management* 181(1–2):77–97.

White, P. J., and R. A. Garrott. 2005. Yellowstone's ungulates after wolves—expectations, realizations, and predictions. *Biological Conservation* 125(2):141–52.

White, Robert. 1997. Librarian's death fuels cull debate. *Arlington Journal*, October 28.

Whitehead, Hal. 2006. Sperm whales in ocean ecosystems. In *Whales, whaling, and ocean ecosystems*, ed. James A. Estes, Douglas P. DeMaster, Daniel F. Doak, Terrie M. Williams, and Robert L. Brownell Jr., 324–34.

Whitehead, Hal, and Randall Reeves. 2005. Killer whales and whaling: The scavenging hypothesis. *Biology Letters* 1:415–18.

Whitney, G. G. 1984. Fifty years of change in the arboreal vegetation of Heart's Content, an old growth hemlock-white pine-northern hardwood stand. *Ecology* 65:403–8.

Wiegmann, Shannon M., and Donald M. Waller. 2006. Fifty years of change in northern upland forest understories: Identity and traits of "winner" and "loser" plant species. *Biological Conservation* 129:109–23.

Wilcove, David S. 1985. Nest predation in forest tracts and the decline of migratory songbirds. *Ecology* 66:1211–14.

———. 1988. Changes in the avifauna of the Great Smoky Mountains: 1947–1983. *Wilson Bulletin* 100:256–71.

———. 1990. Empty skies. *Nature Conservancy*, January–February: 4–13.

Wilcove, D. S., and J. W. Terborgh. 1984. Patterns of population decline in birds. *American Birds* 38:10–13.

Williams, Ted. 1990. Waiting for wolves to howl in Yellowstone. *Audubon*, September–October: 32–41.

———. 2006. They should shoot horses, shouldn't they? *High Country News*, December 11.

Williams, Terrie M. 2006. Reassessing why big fierce animals are rare. Plenary lecture, American Physiological Society, Virginia Beach, Virginia.

Williams, Terrie M., James A. Estes, Daniel F. Doak, and Alan M. Springer. 2004. Killer appetites: Assessing the role of predators in ecological communities. *Ecology* 85(12):3373–84.

Willis, Edwin O. 1974. Populations and local extinctions of birds on Barro Colorado Island, Panama. *Ecological Monographs* 44:153–69.

Wilmers, Christopher C., Robert. L. Crabtree, Douglas W. Smith, Kerry M. Murphy, and Wayne M. Getz. 2003. Trophic facilitation by introduced top predators: Grey wolf subsidies to scavengers in Yellowstone National Park. *Journal of Animal Ecology* 72:909–916.

Wilmers, Christopher C., Daniel R. Stahler, Robert. L. Crabtree, Douglas W. Smith, Kerry M. Murphy, and Wayne M. Getz. 2003. Resource dispersion and consumer dominance: Scavenging at wolf- and hunter-killed carcasses in Greater Yellowstone, USA. *Ecology Letters* 6:996–1003.

Wilson, Edward O. 1978. On human nature. Cambridge, MA: Harvard University Press.

———. 1984. *Biophilia*. Cambridge, MA: Harvard University Press.

———. 1992. *The diversity of life*. Cambridge, MA: Harvard University Press.

Wilson, Paul J., Sonya Grewal, Ian D. Lawford, Jennifer N. M. Heal, Angela G. Granacki, David Pennock, John B. Theberge, Mary T. Theberge, Dennis R. Voight, Will Waddell, Robert E. Chambers, Paul C. Paquet, Gloria Goulet, Dean Cluff, and Bradley N. White. 2000. DNA profiles of the eastern Canadian wolf and the red wolf provide evidence for a common evolutionary history independent of the gray wolf. *Canadian Journal of Zoology* 78:2156–66.

Wilson, P. J., S. Grewal, T. McFadden, R. C. Chambers, and B. N. White. 2003. Mitochondrial DNA extracted from eastern North American wolves killed in the 1800s is not of gray wolf origin. *Canadian Journal of Zoology* 81:936–40.

Winter, Linda. 2006. Impacts of feral and free-ranging cats on bird species of special concern. http://www.abcbirds.org/newsandreports/NFWF.pdf (accessed December 21, 2007).

Wirsing, Aaron J., Michael R. Heithaus, and Lawrence M. Dill. 2007a. Can you dig it? Use of excavation, a risky foraging tactic by dugongs is sensitive to predation danger. *Animal Behaviour* 74:1085–91.

———. 2007b. Fear factor: Do dugongs (*Dugong dugon*) trade food for safety from tiger sharks (*Galeocerdo cuvier*)? *Oecologia* 153:1031–40.

———. 2007c. Living on the edge: Dugongs prefer to forage in microhabitats that allow escape rather than avoidance of predators. *Animal Behaviour.* 74:93–101.

Wolverton, Steve, James H. Kennedy, and John D. Cornelius. 2007. A paleozoological perspective on white-tailed deer (*Odocoileus virginianus texana*) population density and body size in central Texas. *Environmental Management* 39:545–52.

Woodroffe, Rosie. 2001. Strategies for carnivore conservation: Lessons from contemporary extinctions. In *Carnivore Conservation*, ed. John L. Gittleman, Stephan M. Funk, David W. MacDonald, and Robert K. Wayne, 61–92.

Wootton, Timothy J., and Amy L. Downing. 2003. Understanding the effects of reduced biodiversity: A comparison of two approaches. In *The importance of species: Perspectives on expendability and triage*, ed. Peter Kareiva and Simon A. Levin, 85–104.

Wootton, J. T., M. E. Power, R. T. Paine, and C. A. Pfister. 1996. Effects of productivity, consumers, competitors, and El Niño events on food chain patterns in a rocky intertidal community. *Proceedings of the National Academy of Sciences* 93:13855–58.

Worm, Boris, Heike K. Lotze, and Ransom A. Myers. 2006. Ecosystem effects of fishing and whaling in the North Pacific and Atlantic Ocean. In *Whales, whaling, and ocean ecosystems*, ed. James A. Estes, Douglas P. Demaster, Daniel F. Doak, Terrie M. Williams, and Robert L. Brownell Jr., 335–43.

Worster, Donald. 1994. *Nature's economy: A history of ecological ideas.* 2nd ed. Cambridge: Press Syndicate of the University of Cambridge.

Wright, S. Joseph, Matthew E. Gompper, and Bonifacio Deleon. 1994. Are large predators keystone species in neotropical forests? The evidence from Barro Colorado Island. *Oikos* 71:279–94.

Yellowstone National Park. 1997. *Yellowstone's northern range: Complexity and change in a wildland ecosystem.* Mammoth Hot Springs, WY: National Park Service.

Yoon, Carol Kaesuk. 1996. Pronghorn's speed may be legacy of past predators. *New York Times*, December 24.

Young, Stanley P., and Edward A. Goldman. 1944. *The wolves of North America. Part 1: Their history, life habits, economic status, and control.* New York: Dover Publications.

Zakin, Susan. 1996. Back to the Pleistocene. *Sports Afield*, February: 43–44.

Zaradic, Patricia A., and Oliver R. W. Pergams. 2007. Videophilia: Implications for childhood development and conservation. *Journal of Developmental Processes* 2(1): 130–44.

Ziegler, Christian, and Egbert Giles Leigh Jr. 2002. *A magic web.* New York: Oxford University Press.

Zimov, Sergey A. 2005. Pleistocene Park: Return of the mammoth's ecosystem. *Science* 308:796–98.

Zimov, S. A., V. I. Chuprynin, A. P. Oreshko, F. S. Chapin III, J. F. Reynolds, and M. C. Chapin. 1995. Steppe-tundra transition: A herbivore-driven biome shift at the end of the Pleistocene. *American Naturalist* 146:765–94.

Zuberbühler, Klaus, and David Jenny. 2002. Leopard predation and human evolution. *Journal of Human Evolution* 43:873–86.

Zumbo, Jim. 1987. Should we cry wolf? *Outdoor Life*, December: 50, 98–100.

INDEX

A NOTE ON THE AUTHOR

William Stolzenburg has studied predator control techniques and monitored endangered species. He has written hundreds of magazine features and columns on the ecology of rarity and extinction for *Science News* and *Nature Conservancy*, among others. He lives in Shepherdstown, West Virginia.